Mineralogy: An Introduction

Mineralogy: An Introduction

Zack Reed

SYRAWOOD
PUBLISHING HOUSE
New York

Published by Syrawood Publishing House,
750 Third Avenue, 9th Floor,
New York, NY 10017, USA
www.syrawoodpublishinghouse.com

Mineralogy: An Introduction
Zack Reed

International Standard Book Number: 978-1-64740-112-2 (Hardback)

Cataloging-in-Publication Data

Mineralogy : an introduction / Zack Reed.
    p. cm.
Includes bibliographical references and index.
ISBN 978-1-64740-112-2
1. Mineralogy. 2. Minerals. 3. Physical geology. I. Reed, Zack.
QE363.2 .M56 2022
549--dc23

TABLE OF CONTENTS

PREFACE

The purpose of this book is to help students understand the fundamental concepts of this discipline. It is designed to motivate students to learn and prosper. I am grateful for the support of my colleagues. I would also like to acknowledge the encouragement of my family.

The branch of geology that deals with the scientific study of the chemistry, crystal structure and physical properties of minerals is referred to as mineralogy. It also examines mineralized artifacts. It is particularly concerned with the processes related to mineral origin and formation, geographical distribution of minerals, their classification as well as their utilization. The physical properties of minerals are classified on the basis of density, hardness, fracture, radioactivity and solubility. Some of the subdisciplines within this field are optical mineralogy, systematic mineralogy and biomineralogy. Optical mineralogy focuses on the study of minerals and rocks by measuring their optical properties. The classification and identification of minerals falls under the domain of systematic mineralogy. The book aims to shed light on some of the unexplored aspects of mineralogy. While understanding the long-term perspectives of the topics, it makes an effort in highlighting their impact as a modern tool for the growth of the discipline. This book, with its detailed analyses and data, will prove immensely beneficial to professionals and students involved in this area at various levels.

A foreword for all the chapters is provided below:

Chapter –Mineral and its Types

The solid chemical compound which exists naturally in its pure form is called a mineral. Some of the different types of minerals are silicate minerals, oxide minerals and carbonate minerals. This chapter will provide a brief introduction to these types of minerals as well as the different properties of minerals such as chemical properties and electrical properties.

Chapter – Mineralogy and its Branches

The domain of geology which focuses on the scientific study of the crystal structure, chemistry and physical properties of minerals is known as mineralogy. A few of its branches are optical mineralogy, magnetic mineralogy, environmental mineralogy and clay mineralogy. This chapter has been carefully written to provide an easy understanding of these branches of mineralogy.

Chapter – Mineral Analysis and Identification

Some of the ways of identifying a mineral are by analyzing how it reflects light, testing its hardness, identify fractures, etc. A deeper study into the analysis of minerals can be conducted using tools such as electron microscopes. The topics elaborated in this chapter will help in gaining a better perspective about the ways of identifying minerals and analyzing them.

Chapter – Mineral Formation and Defects

Minerals can be formed under a wide range of physical and chemical conditions such as fluctuating temperature and pressure. Defects in mineral crystals are the distortions of the ordered arrangement of atoms in the crystalline lattice. This chapter closely examines the key concepts related to the formation of minerals as well as the structural defects in them.

Chapter – Crystallography

The branch of science which studies the arrangement of atoms within a crystalline solid is known as crystallography. Some of the important areas of study within this field are crystal structure, crystal forms and crystallographic effect. This chapter discusses these focus areas of crystallography in detail.

Zack Reed

Mineral and its Types

1

- **Mineraloid**
- **Properties of Minerals**
- **Silicate Minerals**
- **Oxide Minerals**
- **Carbonate Minerals**

The solid chemical compound which exists naturally in its pure form is called a mineral. Some of the different types of minerals are silicate minerals, oxide minerals and carbonate minerals. This chapter will provide a brief introduction to these types of minerals as well as the different properties of minerals such as chemical properties and electrical properties.

Mineral is a naturally occurring homogeneous solid with a definite chemical composition and a highly ordered atomic arrangement; it is usually formed by inorganic processes. There are several thousand known mineral species, about 100 of which constitute the major mineral components of rocks; these are the so-called rock-forming minerals.

A mineral, which by definition must be formed through natural processes, is distinct from the synthetic equivalents produced in the laboratory. Artificial versions of minerals, including emeralds, sapphires, diamonds, and other valuable gemstones, are regularly produced in industrial and research facilities and are often nearly identical to their natural counterparts.

Amethyst.

By its definition as a homogeneous solid, a mineral is composed of a single solid substance of uniform composition that cannot be physically separated into simpler chemical compounds. Homogeneity is determined relative to the scale on which it is defined. A specimen that appears homogeneous to the unaided eye, for example, may reveal several mineral components under a microscope or upon exposure to X-ray diffraction techniques. Most rocks are composed of several different minerals; e.g., granite consists of feldspar, quartz, mica, and amphibole. In addition, gases and liquids are excluded by a strict interpretation of the above definition of a mineral. Ice, the solid state of water (H_2O), is considered a mineral, but liquid water is not; liquid mercury, though sometimes found in mercury ore deposits, is not classified as a mineral either. Such substances that resemble minerals in chemistry and occurrence are dubbed mineraloids and are included in the general domain of mineralogy.

Since a mineral has a definite composition, it can be expressed by a specific chemical formula. Quartz (silicon dioxide), for instance, is rendered as SiO_2, because the elements silicon (Si) and oxygen (O) are its only constituents and they invariably appear in a 1:2 ratio. The chemical makeup of most minerals is not as well defined as that of quartz, which is a pure substance. Siderite, for example, does not always occur as pure iron carbonate ($FeCO_3$); magnesium (Mg), manganese (Mn), and, to a limited extent, calcium (Ca) may sometimes substitute for the iron. Since the amount of the replacement may vary, the composition of siderite is not fixed and ranges between certain limits, although the ratio of the metal cation to the anionic group remains fixed at 1:1. Its chemical makeup may be expressed by the general formula $(Fe, Mn, Mg, Ca)CO_3$, which reflects the variability of the metal content.

Trigonal system.

Minerals display a highly ordered internal atomic structure that has a regular geometric form. Because of this feature, minerals are classified as crystalline solids. Under favourable conditions, crystalline materials may express their ordered internal framework by a well-developed external form, often referred to as crystal form or morphology. Solids that exhibit no such ordered internal arrangement are termed amorphous. Many amorphous natural solids, such as glass, are categorized as mineraloids.

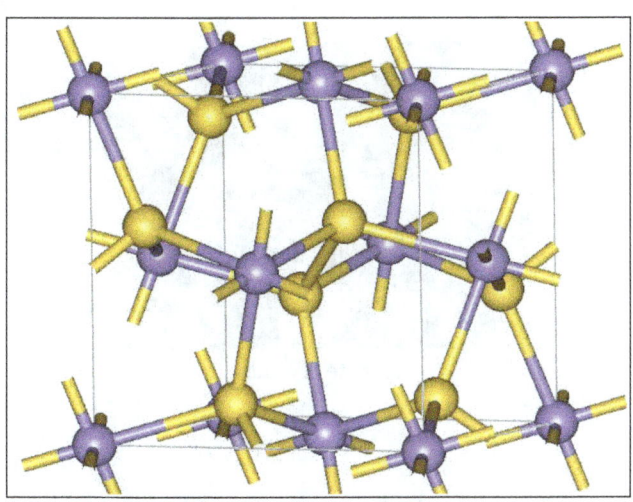

Pyrite.

Traditionally, minerals have been described as resulting exclusively from inorganic processes; however, current mineralogic practice often includes as minerals those compounds that are organically produced but satisfy all other mineral requirements. Aragonite ($CaCO_3$) is an example of an inorganically formed mineral that also has an organically produced, yet otherwise identical, counterpart; the shell (and the pearl, if it is present) of an oyster is composed to a large extent of organically formed aragonite. Minerals also are produced by the human body: hydroxylapatite [$Ca_5(PO_4)_3(OH)$] is the chief component of bones and teeth, and calculi are concretions of mineral substances found in the urinary system.

While minerals are classified in a logical manner according to their major anionic (negatively charged) chemical constituents into groups such as oxides, silicates, and nitrates, they are named in a far less scientific or consistent way. Names may be assigned to reflect a physical or chemical property, such as colour, or they may be derived from various subjects deemed appropriate, such as, for example, a locality, public figure, or mineralogist. Some examples of mineral names and their derivations follow: albite ($NaAlSi_3O_8$) is from the word (albus) for "white" in reference to its colour; goethite ($FeO \cdot OH$) is in honour of Johann Wolfgang von Goethe, the German poet; manganite ($MnO \cdot OH$) reflects the mineral's composition; franklinite ($ZnFe_2O_4$) is named after Franklin, New Jersey, U.S., the site of its occurrence as the dominant ore mineral for zinc (Zn); and sillimanite (Al_2SiO_4) is in honour of the American chemist Benjamin Silliman. Since 1960 the Commission on New Minerals and Mineral Names of the International Mineralogical Association has reviewed descriptions of new minerals and proposals for new mineral names and has attempted to remove inconsistencies. Any new mineral name must be approved by this committee, and the type material is usually stored in a museum or university collection.

Nature of Minerals

Morphology

Nearly all minerals have the internal ordered arrangement of atoms and ions that is the defining characteristic of crystalline solids. Under favourable conditions, minerals may grow as well-formed crystals, characterized by their smooth plane surfaces and regular geometric forms. Development of this good external shape is largely a fortuitous outcome of growth and does not affect the basic properties of a crystal. Therefore, the term crystal is most often used by material scientists to refer to any solid with an ordered internal arrangement, without regard to the presence or absence of external faces.

Azurite crystals.

Symmetry Elements

The external shape, or morphology, of a crystal is perceived as its aesthetic beauty, and its geometry reflects the internal atomic arrangement. The external shape of well-formed crystals expresses the presence or absence of a number of symmetry elements. Such symmetry elements include rotation axes, rotoinversion axes, a centre of symmetry, and mirror planes.

Crystals Cave.

A rotation axis is an imaginary line through a crystal around which it may be rotated and repeat itself in appearance one, two, three, four, or six times during a complete rotation. (For example, a sixfold rotation occurs when the crystal repeats itself each 60°—that is, six times in a 360° rotation.)

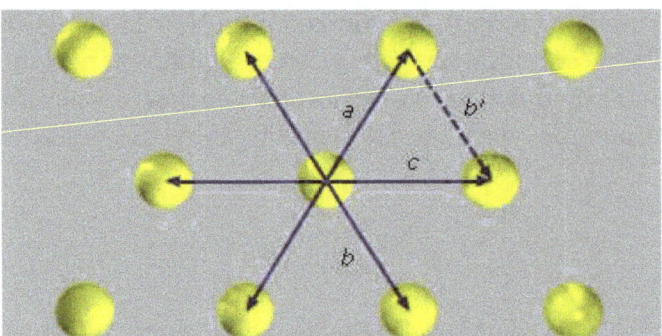

Arrows indicates translational symmetries of the lattice.

A rotoinversion axis combines rotation about an axis of rotation with inversion. Rotoinversion axes are symbolized as $\bar{1}$, $\bar{2}$, $\bar{3}$, $\bar{4}$, and $\bar{6}$, where $\bar{1}$ is equivalent to a centre of symmetry (or inversion), $\bar{2}$ is equivalent to a mirror plane, and 3 is equivalent to a threefold rotation axis plus a centre of symmetry. When the axis of the crystal is vertical, $\overline{4}$ is characterized by two top faces with identical faces upside down underneath. $\overline{6}$ is equivalent to a threefold rotation axis with a mirror plane perpendicular to the axis.

A centre of symmetry exists in a crystal if an imaginary line can be extended from any point on its surface through its centre and a similar point is present along the line equidistant from the centre. This is equivalent to 1, or inversion. There is a relatively simple procedure for recognizing a centre of symmetry in a well-formed crystal. With the crystal laid down on any face on a tabletop, the presence of a face of equal size and shape, but inverted, in a horizontal position at the top of the crystal proves the existence of a centre of symmetry. An imaginary mirror plane (or symmetry plane) can also be used to separate a crystal into halves. In a perfectly developed crystal, the halves are mirror images of one another.

Morphologically, crystals can be grouped into 32 crystal classes that represent the 32 possible symmetry elements and their combinations. These crystal classes, in turn, are grouped into six crystal systems. In decreasing order of overall symmetry content, beginning with the system with the highest and most complex crystal symmetry, they are isometric (or cubic), hexagonal, tetragonal, orthorhombic, monoclinic, and triclinic. (Many sources list seven crystal systems by dividing the hexagonal crystal system into two parts—trigonal and hexagonal.)

The 32 crystal classes and their symmetry contents		
Crystal system	Symmetry content*	Crystal class**
Triclinic	none	1
	i	1
Monoclinic	$1A_2$	2
	1m	m
	i, $1A_2$, 1m	2/m
Orthorhombic	$3A_2$	222
	A_2, 2m	mm2
	i, $3A_2$, 3m	2/m2/m2/m
Tetragonal	$1A_4$	4
	$1A_4$	4
	i, $1A_4$, m	4/m
	$1A_4$, $4A_2$	422
	$1A_4$, 4m	4mm
	$1A_4$, $2A_2$, 2m	42m
	i, $1A_4$, $4A_2$, 5m	4/m2/m2/m
Hexagonal	$1A_3$	3
	$1A_3$ (= i + $1A_3$)	3
	$1A_3$, $3A_2$	32
	$1A_3$, 3m	3m
	$1A_3$, $3A_2$, 3m ($1A_3$ = i + $1A_3$)	32/m
	$1A_6$	6
	$1A_6$ (= $1A_3$ + m)	6
	i, $1A_6$, 1m	6/m

	$1A_6$, $6A_2$	622
	$1A_6$, 6m	6mm
	$1A_6$, $3A_2$, 3m ($1A_6 = 1A_3 + m$)	6m2
	i, $1A_6$, $6A_2$, 7m	6/m2/m2/m
Isometric	$3A_2$, $4A_3$	23
	$3A_2$, 3m, $4A_3$ ($1A_3 = 1A_3 + i$)	2/m3
	$3A_4$, $4A_3$, $6A_2$	432
	$3A_4$, $4A_3$, 6m	43m
	$3A_4$, $4A_3$, $6A_2$, 9m ($1A_3 = 1A_3 + i$)	4/m32/m
*Abbreviations used in column 2: i = inversion (or centre of symmetry); A = axis of rotation; A_2 = axis of twofold rotation; A_3 = axis of threefold rotation; A_4 = axis of fourfold rotation; and A_6 = axis of sixfold rotation; A = axis of rotoinversion; A_3 = axis of threefold rotoinversion; A_4 = axis of fourfold rotoinversion; A_6 = axis of sixfold rotoinversion; m = mirror, or symmetry, plane.		
**Symbolic representation used in column 3: rotation axes are shown as 1, 2, 3, 4, or 6 (in which 2 = twofold rotation, 3 = threefold rotation, etc.); rotoinversion axes are shown as 3, 4, or 6 (in which 3 is a threefold rotoinversion axis, etc.); centre of symmetry i is equivalent to 1; mirrors are represented by m; rotation axes perpendicular to mirror planes are shown by the notation 2/m, 4/m, or 6/m, in which 2/m is a twofold axis perpendicular to a mirror, etc.		

The systems may be described in terms of crystallographic axes used for reference. The c axis is normally the vertical axis. The isometric system exhibits three mutually perpendicular axes of equal length (a_1, a_2, and a_3). The orthorhombic and tetragonal systems also contain three mutually perpendicular axes; in the former system all the axes are of different lengths (a, b, and c), and in the latter system two axes are of equal length (a_1 and a_2) while the third (vertical) axis is either longer or shorter (c). The hexagonal system contains four axes: three equal-length axes (a_1, a_2, and a_3) intersect one another at 120° and lie in a plane that is perpendicular to the fourth (vertical) axis of a different length. Three axes of different lengths (a, b, and c) are present in both the monoclinic and triclinic systems. In the monoclinic system, two axes intersect one another at an oblique angle and lie in a plane perpendicular to the third axis; in the triclinic system, all axes intersect at oblique angles.

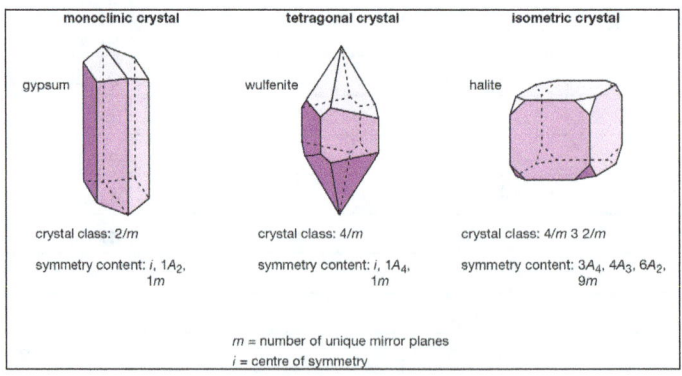

Examples of well-shaped crystals.

Twinning

If two or more crystals form a symmetrical intergrowth, they are referred to as twinned crystals. A new symmetry operation (called a twin element), which is lacking in a single untwinned crystal, relates the individual crystals in a twinned position. There are three twin elements that may relate the crystals of a twin: (1) reflection by a mirror plane (twin plane), (2) rotation about a crystal direction common to both (twin axis) with the angular rotation typically 180°, and (3) inversion about a point (twin centre). An instance of twinning is defined by a twin law that specifies the presence of a plane, an axis, or a centre of twinning. If a twin has three or more parts, it is referred to as a multiple, or repeated, twin.

Internal Structure

Examining Crystal Structures

The external morphology of a mineral is an expression of the fundamental internal architecture of a crystalline substance—i.e., its crystal structure. The crystal structure is the three-dimensional, regular (or ordered) arrangement of chemical units (atoms, ions, and anionic groups in inorganic materials; molecules in organic substances); these chemical units (referred to here as motifs) are repeated by various translational and symmetry operations. The morphology of crystals can be studied with the unaided eye in large well-developed crystals and has been historically examined in considerable detail by optical measurements of smaller well-formed crystals through the use of optical goniometers (instruments that measure the angles between crystal faces). The internal structure of crystalline materials, however, is revealed by a combination of X-ray, neutron, and electron diffraction techniques, supplemented by a variety of spectroscopic methods, including infrared, optical, Mössbauer, and resonance techniques. These methods, used singly or in combination, provide a quantitative three-dimensional reconstruction of the location of the atoms (or ions), the chemical bond types and their positions, and the overall internal symmetry of the structure. The repeat distances in most inorganic structures and many of the atomic and ionic motif sizes are on the order of 1 to 10 angstroms (Å; 1 Å is equivalent to 10^{-8} cm or 3.94×10^{-9} inch) or 10 to 100 nanometres (nm; 1 nm is equivalent to 10^{-7} cm or 10 Å).

Space Groups

Symmetry elements that are observable in the external morphology of crystals, such as rotation and rotoinversion axes, mirror planes, and a centre of symmetry, also are present in their internal atomic structure. In addition to these symmetry elements, there are translations and symmetry operations combined with translations. (Translation is the operation in which a motif is repeated in a linear pattern at intervals that are equal to the translation distance [commonly on the 1 to 10 Å level].) Two examples of translational symmetry elements are screw axes (combining rotation and translation) and glide planes (combining mirroring and translation). The internal translation distances are exceedingly small and can be seen directly only by very high-magnification electron beam techniques, as used in a transmission electron microscope, at magnifications of about 600,000×.

When all possible combinations of translational elements compatible with the 32 crystal classes (also known as point groups) are considered, one arrives at 230 possible ways in which translations, translational symmetry elements (screw axes and glide planes), and translation-free symmetry elements (rotation and rotoinversion axes and mirror planes) can be combined. These translation and symmetry groupings are known as the 230 space groups, representing the various ways in which motifs can be arranged in an ordered three-dimensional array. The symbolic representation of space groups is closely related to that of the Hermann-Mauguin notation of point groups.

Illustrating Crystal Structures

The external morphology of three-dimensional arrangement of crystal structures may be presented on a two-dimensional page or within a computer simulation. Another common method of illustration involves projecting the crystal structure onto a planar surface. The high-temperature form of silicon dioxide (SiO_2) known as tridymite may be represented this way; however, the structural motif units in this case are SiO_4 tetrahedrons composed of a silicon atom surrounded by four oxygen atoms. To further aid the visualization of complex crystal structures within the physical world, three-dimensional physical models of such structures can be built or obtained commercially. Models of this sort reproduce the internal atomic arrangement on an enormously enlarged scale (e.g., one angstrom might be represented by one centimetre).

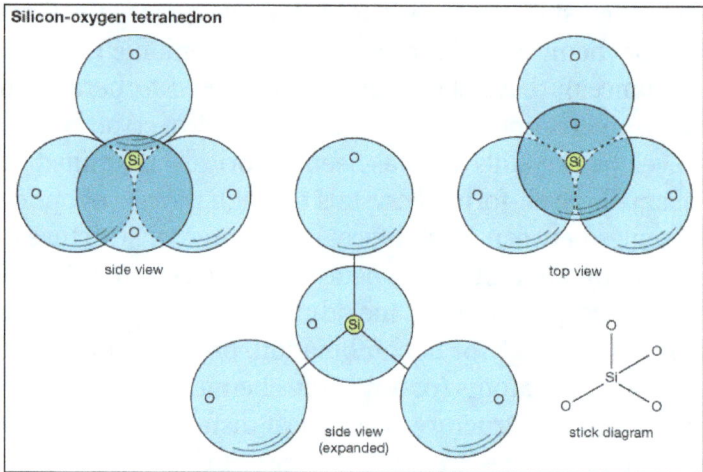

Silicon-oxygen tetrahedrons

Polymorphism

Pyrite

Polymorphism is the ability of a specific chemical composition to crystallize in more than one form. This generally occurs as a response to changes in temperature or pressure or both. The different structures of such a chemical substance are called polymorphic forms, or polymorphs. For example, the element carbon (C) occurs in nature in two different polymorphic forms, depending on the external (pressure and temperature) conditions. These forms are graphite, with a hexagonal structure, and diamond, with an isometric structure. The composition FeS_2 occurs most commonly as pyrite, with an isometric structure, but it is also found as marcasite, which has an orthorhombic internal arrangement. The composition SiO_2 is found in a large number of polymorphs, among them quartz, tridymite, cristobalite, coesite, and stishovite. The stability field (conditions under which a mineral is stable) of these SiO_2 polymorphs can be expressed in a stability diagram, with the external parameters of temperature and pressure as the two axes. In the general quartz field, there is additional polymorphism leading to the notation of high quartz and low quartz, each form having a slightly different internal structure. Cristobalite and tridymite are the high-temperature forms of SiO_2, and indeed these SiO_2 polymorphs occur in high-temperature lava flows. The high-pressure forms of SiO_2 are coesite and stishovite, and these can be found in meteorite craters, formed as a result of high explosive pressures upon quartz-rich sandstones, and in very deep-seated rock formations, as from Earth's upper mantle or very deep in subduction zones.

Chemical Composition

The chemical composition of a mineral is of fundamental importance because its properties greatly depend on it. Such properties, however, are determined not only by the chemical composition but also by the geometry of the constituent atoms and ions and by the nature of the electrical forces that bind them. Thus, for a complete understanding of minerals, their internal structure, chemistry, and bond types must be considered.

Various analytical techniques may be employed to obtain the chemical composition of a mineral. Quantitative chemical analyses mainly use so-called wet analytical methods (e.g., dissolution in acid, flame tests, and other classic techniques of bench chemistry that rely on observation), in which the mineral sample is first dissolved. Various compounds are then precipitated from the solution, which are weighed to obtain a gravimetric analysis. A number of analytical procedures have been introduced that provide faster but somewhat less accurate results. Most analyses use instrumental methods such as optical emission, X-ray fluorescence, atomic absorption spectroscopy, and electron microprobe analysis. Relatively well-established error ranges have been documented for these methods, and samples must be prepared in a specific manner for each technique. A distinct advantage of wet analytical procedures is that they make it possible to determine quantitatively the oxidation states of positively charged atoms, called cations (e.g., Fe^{2+} versus Fe^{3+}), and to ascertain the amount of water in hydrous minerals. It is more difficult to provide this type of information with instrumental techniques.

To ensure an accurate chemical analysis, the selected sample, which might include several minerals, is often made into a thin section (a section of rock less than 1 mm thick cemented for study between clear glass plates). To reduce the effect of the impurities, an instrumental technique, such as electron microprobe analysis, is commonly employed. In this method, quantitative analysis in situ may be performed on mineral grains only 1 micrometre (10^{-4} centimetre) in diameter.

Mineral formulas

Elements may exist in the native (uncombined) state, in which case their formulas are simply their chemical symbols: gold (Au), carbon (C) in its polymorphic form of diamond, and sulfur (S) are common examples. Most minerals, however, occur as compounds consisting of two or more elements; their formulas are obtained from quantitative chemical analyses and indicate the relative proportions of the constituent elements. The formula of sphalerite, ZnS, reflects a one-to-one ratio between atoms of zinc and those of sulfur. In bornite (Cu_5FeS_4), there are five atoms of copper (Cu), one atom of iron (Fe), and four atoms of sulfur. There exist relatively few minerals with constant composition; notable examples include quartz (SiO_2) and kyanite (Al_2SiO_5). Minerals of this sort are termed pure substances. Most minerals display considerable variation in the ions that occupy specific atomic sites within their structure. For example, the iron content of rhodochrosite ($MnCO_3$) may vary over a wide range. As ferrous iron (Fe^{2+}) substitutes for manganese cations (Mn^{2+}) in the rhodochrosite structure, the formula for the mineral might be given in more general terms—namely, $(Mn, Fe)CO_3$. The amounts of manganese and iron are variable, but the ratio of the cation to the negatively charged anionic group remains fixed at one Mn^{2+} or Fe^{2+} atom to one CO_3 group.

Sphalerite

Compositional Variation

As stated above, most minerals exhibit a considerable range in chemical composition. Such variation results from the replacement of one ion or ionic group by another in a particular structure. This phenomenon is termed ionic substitution, or solid solution. Three types of solid solution are possible, and these may be described in terms of their corresponding mechanisms—namely, substitutional, interstitial, and omission.

Substitutional solid solution is the most common variety. For example, as described above, in the carbonate mineral rhodochrosite ($MnCO_3$), Fe^{2+} may substitute for Mn^{2+} in its atomic site in the structure.

The degree of substitution may be influenced by various factors, with the size of the ion being the most important. Ions of two different elements can freely replace one another only if their ionic radii differ by approximately 15 percent or less. Limited substitution can occur if the radii differ by 15 to 30 percent, and a difference of more than 30 percent makes substitution unlikely. These limits, calculated from empirical data, are only approximate.

The temperature at which crystals grow also plays a significant role in determining the extent of ionic substitution. The higher the temperature, the more extensive is the thermal disorder in the crystal structure and the less exacting are the spatial requirements. As a result, ionic substitution that could not have occurred in crystals grown at low temperatures may be present in those grown at higher ones. The high-temperature form of $KAlSi_3O_8$ (sanidine), for example, can accommodate more sodium (Na) in place of potassium (K) than can microcline, its low-temperature counterpart.

An additional factor affecting ionic substitution is the maintenance of a balance between the positive and negative charges in the structure. Replacement of a monovalent ion (e.g., Na+, a sodium cation) by a divalent ion (e.g., Ca^{2+}, a calcium cation) requires further substitutions to keep the structure electrically neutral.

Simple cationic or anionic substitutions are the most basic types of substitutional solid solution. A simple cationic substitution can be represented in a compound of the general form A^+X^- in which cation B^+ replaces in part or in total cation A^+. Both cations in this example have the same valence (+1), as in the substitution of K^+ (potassium ions) for Na^+ (sodium ions) in the NaCl (sodium chloride) structure. Similarly, the substitution of anion X^- by Y^- in an A^+X^- compound represents a simple anionic substitution; this is exemplified by the replacement of Cl^- (chlorine ions) with Br^- (bromine ions) in the structure of KCl (potassium chloride). A complete solid-solution series involves the substitution in one or more atomic sites of one element for another that ranges over all possible compositions and is defined in terms of two end-members. For example, the two end-members of olivine [$(Mg, Fe)_2SiO_4$], forsterite (Mg_2SiO_4) and fayalite (Fe_2SiO_4), define a complete solid-solution series (called the forsterite-fayalite series) in which magnesium cations (Mg^{2+}) are replaced partially or totally by Fe^{2+}.

In some instances, a cation B^{3+} may replace some A^{2+} of compound $A^{2+}X^{2-}$. So that the compound will remain neutral, an equal amount of A^{2+} must concurrently be replaced by a third cation, C^+. This is given in equation form as $2A^{2+} \leftrightarrow B^{3+} + C^+$; the positive charge on each side is the same. Substitutions such as this are termed coupled substitutions. The plagioclase feldspar series exhibits complete solid solution, in the form of coupled substitutions, between its two end-members, albite ($NaAlSi_3O_8$) and anorthite ($CaAl_2Si_2O_8$). Every atomic substitution of $Na+$ by Ca^{2+} is accompanied by the replacement of a silicon cation (Si^{4+}) by an aluminum cation (Al^{3+}), thereby maintaining electrical neutrality: $Na+ + Si^{4+} \leftrightarrow Ca^{2+} + Al^{3+}$.

The second major type of ionic substitution is interstitial solid solution, or interstitial substitution. It takes place when atoms, ions, or molecules fill the interstices (voids) found between the atoms, ions, or ionic groups of a crystal structure. The interstices may take the form of channel-like cavities in certain crystals, such as the ring silicate beryl ($Be_3Al_2Si_6O_{18}$). Potassium, rubidium (Rb), cesium (Cs), and water, as well as helium (He), are some of the large ions and gases found in the tubular voids of beryl.

The least common type of solid solution is omission solid solution, in which a crystal contains one or more atomic sites that are not completely filled. The best-known example is exhibited by pyrrhotite ($Fe_{1-x}S$). In this mineral, each iron atom is surrounded by six neighbouring sulfur atoms. If every iron site in pyrrhotite were occupied by ferrous iron, its formula would be FeS. There are, however, varying percentages of vacancy in the iron site, so that the formula is given as Fe_6S_7 through $Fe_{11}S_{12}$, the latter being very near to pure FeS. The formula for pyrrhotite is normally written as $Fe_{1-x}S$, with x ranging from 0 to 0.2. It is one of the minerals referred to as a defect structure, because it has a structural site that is not completely occupied.

Chemical Bonding

Electrical forces are responsible for the chemical bonding of atoms, ions, and ionic groups that constitute crystalline solids. The physical and chemical properties of minerals are attributable for the most part to the types and strengths of these binding forces; hardness, cleavage, fusibility, electrical and thermal conductivity, and the coefficient of thermal expansion are examples of such properties. On the whole, the hardness and melting point of a crystal increase proportionally with the strength of the bond, while its coefficient of thermal expansion decreases. The extremely strong forces that link the carbon atoms of diamond, for instance, are responsible for its distinct hardness. Periclase (MgO) and halite (NaCl) have similar structures; however, periclase has a melting point of 2,800 °C

(5,072 °F) whereas halite melts at 801 °C (1,474 °F). This discrepancy reflects the difference in the bond strength of the two minerals: since the atoms of periclase are joined by a stronger electrical force, a greater amount of heat is needed to separate them.

The electrical forces, called chemical bonds, can be divided into five types: ionic, covalent, metallic, van der Waals, and hydrogen bonds. Classification in this manner is largely one of expediency; the chemical bonds in a given mineral may in fact possess characteristics of more than one bond type. For example, the forces that link the silicon and oxygen atoms in quartz exhibit in nearly equal amount the characteristics of both ionic and covalent bonds. As stated above, the electrical interaction between the atoms of a crystal determine its physical and chemical properties. Thus, classifying minerals according to their electrical forces will cause those species with similar properties to be grouped together. This fact justifies classification by bond type.

Ionic Bonds

Atoms have a tendency to gain or lose electrons so that their outer orbitals become stable; this is normally accomplished by these orbitals being filled with the maximum allowed number of valence electrons. Metallic sodium, for example, has one valence electron in its outer orbital; it becomes ionized by readily losing this electron and exists as the cation Na^+. Conversely, chlorine gains an electron to complete its outer orbital, thereby forming the anion Cl^-. In the mineral halite, NaCl (common, or rock, salt), the chemical bonding that holds the Na^+ and Cl^- ions together is the attraction between the two opposite charges. This bonding mechanism is referred to as ionic, or electrovalent.

Crystal Bonding

Ionically bonded crystals typically display moderate hardness and specific gravity, rather high melting points, and poor thermal and electrical conductivity. The electrostatic charge of an ion is evenly distributed over its surface, and so a cation tends to become surrounded with the maximum number of anions that can be arranged around it. Since ionic bonding is nondirectional, crystals bonded in this manner normally display high symmetry.

Covalent Bonds

In the discussion of the ionic bond, it was noted that chlorine readily gains an electron to achieve a stable electron configuration. An incomplete outer orbital places a chlorine atom in a highly

reactive state, so it attempts to combine with nearly any atom in its proximity. Because its closest neighbour is usually another chlorine atom, the two may bond together by sharing one pair of electrons. As a result of this extremely strong bond, each chlorine atom enters a stable state.

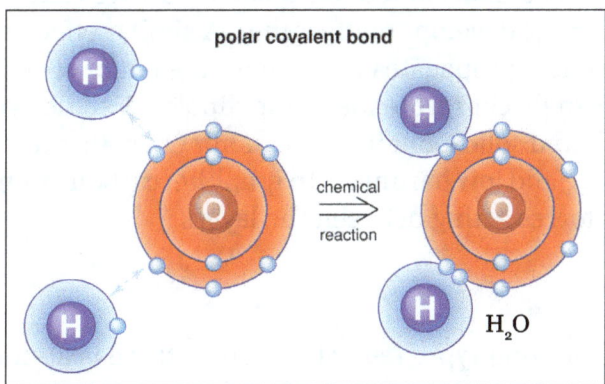

Polar covalent bond

The electron-sharing, or covalent, bond is the strongest of all chemical bond types. Minerals bonded in this manner display general insolubility, great stability, and a high melting point. Crystals of covalently bonded minerals tend to exhibit lower symmetry than their ionic counterparts because the covalent bond is highly directional, localized in the vicinity of the shared electrons.

The Cl_2 molecules formed by linking two neighbouring chlorine atoms are stable and do not combine with other molecules. Atoms of some elements, however, have more than one electron in the outer orbital and thus may bond to several neighbouring atoms to form groups, which in turn may join together in larger combinations. Carbon, in the polymorphic form of diamond, is a good example of this type of covalent bonding. There are four valence electrons in a carbon atom, so that each atom bonds with four others in a stable tetrahedral configuration. A continuous network is formed by the linkage of every carbon atom in this manner. The rigid diamond structure results from the strong localization of the bond energy in the vicinity of the shared electrons; this makes diamond the hardest of all natural substances. Diamond does not conduct electricity, because all the valence electrons of its constituent atoms are shared to form bonds and therefore are not mobile.

Metallic Bonds

Bonding in metals is distinct from that in their salts, as reflected in the significant differences between the properties of the two groups. In contrast to salts, metals display high plasticity, tenacity, ductility, and conductivity. Many are characterized by lower hardness and have higher melting and boiling points than, for example, covalently bonded materials. All these properties result from a metallic bonding mechanism that can be envisioned as a collection of positively charged ions immersed in a cloud of valence electrons. The attraction between the cations and the electrons holds a crystal together. The electrons are not bound to any particular cation and are thus free to move throughout the structure. In fact, in the metals sodium, cesium, rubidium, and potassium, the radiant energy of light can cause electrons to be removed from their surfaces entirely. (This result is known as the photoelectric effect.) Electron mobility is responsible for the ability of metals to conduct heat and electricity. The native metals are the only minerals to exhibit pure metallic bonding.

Van Der Waals Bonds

Neutral molecules may be held together by a weak electric force known as the van der Waals bond. It results from the distortion of a molecule so that a small positive charge develops on one end and a corresponding negative charge develops on the other. A similar effect is induced in neighbouring molecules, and this dipole effect propagates throughout the entire structure. An attractive force is then formed between oppositely charged ends of the dipoles. Van der Waals bonding is common in gases and organic liquids and solids, but it is rare in minerals. Its presence in a mineral defines a weak area with good cleavage and low hardness. In graphite, carbon atoms lie in covalently bonded sheets with van der Waals forces acting between the layers.

Hydrogen Bonds

In addition to the four major bond types described above, there is an interaction called hydrogen bonding. This takes place when a hydrogen atom, bonded to an electronegative atom such as oxygen, fluorine, or nitrogen, is also attracted to the negative end of a neighbouring molecule. A strong dipole-dipole interaction is produced, forming a bond between the two molecules. Hydrogen bonding is common in hydroxides and in many of the layer silicates—e.g., micas and clay minerals.

Classification of Minerals

Since the middle of the 19th century, minerals have been classified on the basis of their chemical composition. Under this scheme, they are divided into classes according to their dominant anion or anionic group (e.g., halides, oxides, and sulfides). Several reasons justify use of this criterion as the distinguishing factor at the highest level of mineral classification. First, the similarities in properties of minerals with identical anionic groups are generally more pronounced than those with the same dominant cation. For example, carbonates have stronger resemblance to one another than do copper minerals. Secondly, minerals that have identical dominant anions are likely to be found in the same or similar geologic environments. Therefore, sulfides tend to occur together in vein or replacement deposits, while silicate-bearing rocks make up much of Earth's crust. Third, current chemical practice employs a nomenclature and classification scheme for inorganic compounds based on similar principles.

Investigators have found, however, that chemical composition alone is insufficient for classifying minerals. Determination of internal structures, accomplished through the use of X rays, allows a more complete appreciation of the nature of minerals. Chemical composition and internal structure together constitute the essence of a mineral and determine its physical properties; thus, classification should rely on both. Crystallochemical principles—i.e., those relating to both chemical composition and crystal structure—were first applied by the British physicist W. Lawrence Bragg and the Norwegian mineralogist Victor Moritz Goldschmidt in the study of silicate minerals. The silicate group was subdivided in part on the basis of composition but mainly according to internal structure. Based on the topology of the SiO_4 tetrahedrons, the subclasses include framework, chain, and sheet silicates, among others. Such mineral classifications are logical and well-defined.

The broadest divisions of the classification used in the present discussion are:

1. Native elements,

2. Sulfides,

3. Sulfosalts,

4. Oxides and hydroxides,

5. Halides,

6. Carbonates,

7. Nitrates,

8. Borates,

9. Sulfates,

10. Phosphates, and

11. Silicates.

Native Elements

Apart from the free gases in Earth's atmosphere, some 20 elements occur in nature in a pure (i.e., uncombined) or nearly pure form. Known as the native elements, they are partitioned into three families: metals, semimetals, and nonmetals. The most common native metals, which are characterized by simple crystal structures, make up three groups: the gold group, consisting of gold, silver, copper, and lead; the platinum group, composed of platinum, palladium, iridium, and osmium; and the iron group, containing iron and nickel-iron. Mercury, tantalum, tin, and zinc are other metals that have been found in the native state. The native semimetals are divided into two isostructural groups (those whose members share a common structure type): (1) antimony, arsenic, and bismuth, with the latter two being more common in nature, and (2) the rather uncommon selenium and tellurium. Carbon, in the form of diamond and graphite, and sulfur are the most important native nonmetals.

Native elements	
Metals	
Gold group	
Gold	Au
Silver	Ag
Copper	Cu
Platinum group	
Platinum	Pt
Iron group	
Iron	Fe
(Kamacite)	(Fe, Ni)
(Taenite)	(Fe, Ni)
Semimetals	
Arsenic group	
Arsenic	As

Bismuth	Bi
Nonmetals	
Sulfur	S
Diamond	C
Graphite	C

Metals

Gold, silver, and copper are members of the same group (column) in the periodic table of elements and therefore have similar chemical properties. In the uncombined state, their atoms are joined by the fairly weak metallic bond. These minerals share a common structure type, and their atoms are positioned in a simple cubic closest-packed arrangement. Gold and silver both have an atomic radius of 1.44 angstroms (Å), or 1.44×10^{-7} millimetre, which enables complete solid solution to take place between them. The radius of copper is significantly smaller (1.28 Å), and as such copper substitutes only to a limited extent in gold and silver. Likewise, native copper contains only trace amounts of gold and silver in its structure.

Ore

Because of their similar crystal structure, the members of the gold group display similar physical properties. All are rather soft, ductile (capable of being drawn into wire), malleable (capable of being shaped by a hammer or rollers), and sectile (capable of being cut smoothly by a knife or other instrument); gold, silver, and copper serve as excellent conductors of electricity and heat and exhibit metallic lustre and hackly fracture (a type of fracture characterized by sharp jagged surfaces). These properties are attributable to their metallic bonding. The gold-group minerals crystallize in the isometric system and have high densities as a consequence of cubic closest packing.

In addition to the elements listed above, the platinum group also includes rare mineral alloys such as iridosmine. The members of this group are harder than the metals of the gold group and also have higher melting points.

The iron-group metals are isometric and have a simple cubic packed structure. Its members include pure iron, which is rarely found on the surface of Earth, and two species of nickel-iron (kamacite and taenite), which have been identified as common constituents of meteorites. Native iron has been found in basalts of Disko Island, Greenland and nickel-iron in Josephine and Jackson counties, Oregon. The atomic radii of iron and nickel are both approximately 1.24 Å, and so nickel is a frequent substitute for iron. Earth's core is thought to be composed largely of such iron-nickel alloys.

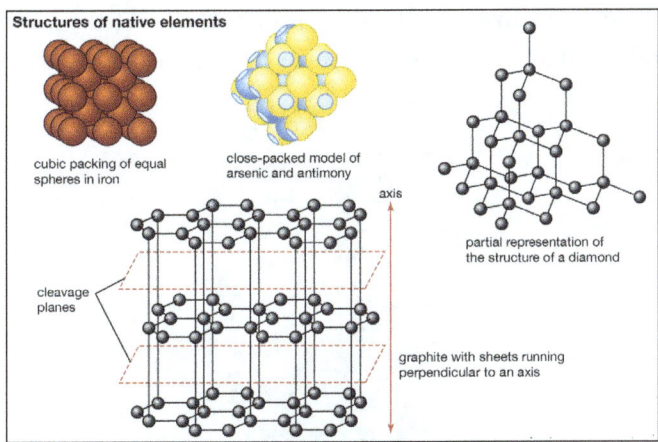

The image above depicts the structures of some native elements. (A) Close-packed model of simple cubic packing of equal spheres, as shown by iron. Each sphere is surrounded by eight closest neighbours. (B) Close-packed model of the structure of arsenic and antimony. Flat areas represent overlap between adjoining atoms. (C) Partial representation of the structure of diamond. (D) The structure of graphite with sheets perpendicular to the *c* axis.

Semimetals

The semimetals antimony, arsenic, and bismuth have a structure type distinct from the simple-packed spheres of the metals. In these semimetals, each atom is positioned closer to three of its neighbouring atoms than to the rest. The structure of antimony and arsenic is composed of spheres that intersect along flat circular areas.

Arsenic (gray) with realgar (red) and orpiment (yellow).

The covalent character of the bonds joining the four closest atoms is linked to the electronegative nature of the semimetals, reflected by their position in the periodic table. Members of this group are fairly brittle, and they do not conduct heat and electricity nearly as well as the native metals. The bond type suggested by these properties is intermediate between metallic and covalent; it is consequently stronger and more directional than pure metallic bonding, resulting in crystals of lower symmetry.

Nonmetals

The native nonmetals diamond, fullerene, graphite, and sulfur are structurally distinct from the metals and semimetals. The structure of sulfur (atomic radius = 1.04 Å), usually orthorhombic in form, may contain limited solid solution by selenium (atomic radius = 1.16 Å).

The Hope diamond; in the Smithsonian Institution, Washington, D.C.

The polymorphs of carbon—graphite, fullerene, and diamond—display dissimilar structures, resulting in their differences in hardness and specific gravity. In diamond, each carbon atom is bonded covalently in a tetrahedral arrangement, producing a strongly bonded and exceedingly close-knit but not closest-packed structure. The carbon atoms of graphite, however, are arranged in six-membered rings in which each atom is surrounded by three close-by neighbours located at the vertices of an equilateral triangle. The rings are linked to form sheets, called graphene, that are separated by a distance exceeding one atomic diameter. Van der Waals forces act perpendicular to the sheets, offering a weak bond, which, in combination with the wide spacing, leads to perfect basal cleavage and easy gliding along the sheets. Fullerenes are found in meta-anthracite, in fulgurites, and in clays from the Cretaceous-Tertiary boundary in New Zealand, Spain, and Turkmenistan as well as in organic-rich layers near the Sudbury nickel mine of Canada.

Sulfides

This important class includes most of the ore minerals. The similar but rarer sulfarsenides are grouped here as well. Sulfide minerals consist of one or more metals combined with sulfur; sulfarsenides contain arsenic replacing some of the sulfur.

Sulfides are generally opaque and exhibit distinguishing colours and streaks. (Streak is the colour of a mineral's powder.) The nonopaque varieties (e.g., cinnabar, realgar, and orpiment) possess high refractive indices, transmitting light only on the thin edges of a specimen.

Few broad generalizations can be made about the structures of sulfides, although these minerals can be classified into smaller groups according to similarities in structure. Ionic and covalent bonding are found in many sulfides, while metallic bonding is apparent in others as evidenced by their metal properties. The simplest and most symmetric sulfide structure is based on the architecture of the sodium chloride structure. A common sulfide mineral that crystallizes in this manner is the ore mineral of lead, galena. Its highly symmetric form consists of cubes modified by octahedral faces at their corners. The structure of the common sulfide pyrite (FeS_2) also is modeled after the sodium chloride type; a disulfide grouping is located in a position of coordination with six surrounding ferrous iron atoms. The high symmetry of this structure is reflected in the external morphology of pyrite. In another sulfide structure, sphalerite (ZnS), each zinc atom is surrounded by four sulfur atoms in a tetrahedral coordinating arrangement. In a derivative of this structure type, the chalcopyrite ($CuFeS_2$) structure, copper and iron ions can be thought of as having been regularly substituted in the zinc positions of the original sphalerite atomic arrangement.

Arsenopyrite (FeAsS) is a common sulfarsenide that occurs in many ore deposits. It is the chief source of the element arsenic.

Sulfosalts

There are approximately 100 species constituting the rather large and very diverse sulfosalt class of minerals. The sulfosalts differ notably from the sulfides and sulfarsenides with regard to the role of semimetals, such as arsenic (As) and antimony (Sb), in their structures. In the sulfarsenides, the semimetals substitute for some of the sulfur in the structure, while in the sulfosalts they are found instead in the metal site. For example, in the sulfarsenide arsenopyrite (FeAsS), the arsenic replaces sulfur in a marcasite- (FeS_2) type structure. In contrast, the sulfosalt enargite (Cu_3AsS_4) contains arsenic in the metal position, coordinated to four sulfur atoms. A sulfosalt such as Cu_3AsS_4 may also be thought of as a double sulfide, $3Cu_2S \cdot As_2S_5$.

Oxides and Hydroxides

These classes consist of oxygen-bearing minerals; the oxides combine oxygen with one or more metals, while the hydroxides are characterized by hydroxyl $(OH)^-$ groups.

The oxides are further divided into two main types: simple and multiple. Simple oxides contain a single metal combined with oxygen in one of several possible metal:oxygen ratios (X:O): XO, X_2O, X_2O_3, etc. Ice, H_2O, is a simple oxide of the X_2O type that incorporates hydrogen as the cation. Although SiO_2 (quartz and its polymorphs) is the most commonly occurring oxide on silicates because its structure more closely resembles that of other silicon-oxygen compounds. Two nonequivalent metal sites (X and Y) characterize multiple oxides, which have the form XY_2O_4.

Unlike the minerals of the sulfide class, which exhibit ionic, covalent, and metallic bonding, oxide minerals generally display strong ionic bonding. They are relatively hard, dense, and refractory.

Oxides generally occur in small amounts in igneous and metamorphic rocks and also as preexisting grains in sedimentary rocks. Several oxides have great economic value, including the principal ores of iron (hematite and magnetite), chromium (chromite), manganese (pyrolusite, as well as the hydroxides, manganite and romanechite), tin (cassiterite), and uranium (uraninite).

Members of the hematite group are of the X_2O_3 type and have structures based on hexagonal closest packing of the oxygen atoms with octahedrally coordinated (surrounded by and bonded to six atoms) cations between them. Corundum and hematite share a common hexagonal architecture. In the ilmenite structure, iron and titanium occupy alternate Fe-O and Ti-O layers.

The XO_2-type oxides are divided into two groups. The first structure type, exemplified by rutile, contains cations in octahedral coordination with oxygen. The second resembles fluorite (CaF_2); each oxygen is bonded to four cations located at the corners of a fairly regular tetrahedron, and each cation lies within a cube at whose corners are eight oxygen atoms. This latter structure is exhibited by uranium, thorium, and cerium oxides, whose considerable importance arises from their roles in nuclear chemistry.

The spinel-group minerals have type XY_2O_4 and contain oxygen atoms in approximate cubic closest packing. The cations located within the oxygen framework are octahedrally (sixfold) and tetrahedrally (fourfold) coordinated with oxygen.

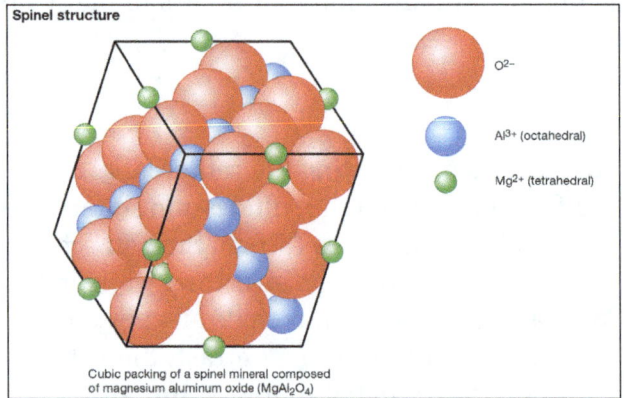

An oxygen layer in the spinel ($MgAl_2O_4$) structure. The large circles represent oxygen in approximate cubic closest packing; the cation layers on each side of the oxygen layer are also shown.

The $(OH)^-$ group of the hydroxides generally results in structures with lower bond strengths than in the oxide minerals. The hydroxide minerals tend to be less dense than the oxides and also are not as hard. All hydroxides form at low temperatures and are found predominantly as weathering products, as, for example, from alteration in hydrothermal veins. Some common hydroxides are brucite [$Mg(OH)_2$], manganite [$MnO{\cdot}OH$], diaspore [$\alpha\text{-}AlO{\cdot}OH$], and goethite [$\alpha\text{-}FeO{\cdot}OH$]. The ore of aluminum, bauxite, consists of a mixture of diaspore, boehmite ($\gamma\text{-}AlO{\cdot}OH$—a polymorph of diaspore), and gibbsite [$Al(OH)_3$], plus iron oxides. Goethite is a common alteration product of iron-rich occurrences and is an iron ore in some localities.

Halides

Members of this class are distinguished by the large-sized anions of the halogens chlorine, bromine, iodine, and fluorine. The ions carry an electric charge of negative one and easily become distorted in the presence of strongly charged bodies. When associated with rather large, weakly polarizing cations of low charge, such as those of the alkali metals, both anions and cations take the form of nearly perfect spheres. Structures composed of these spheres exhibit the highest possible symmetry.

Pure ionic bonding is exemplified best in the isometric halides, for each spherical ion distributes its weak electrostatic charge over its entire surface. These halides manifest relatively low hardness and moderate-to-high melting points. In the solid state they are poor thermal and electric conductors, but when molten they conduct electricity well.

Halogen ions may also combine with smaller, more strongly polarizing cations than the alkali metal ions. Lower symmetry and a higher degree of covalent bonding prevail in these structures. Water and hydroxyl ions may enter the structure, as in atacamite [$Cu_2Cl(OH)_3$].

The halides consist of about 80 chemically related minerals with diverse structures and widely varied origins. The most common are halite (NaCl), sylvite (KCl), chlorargyrite (AgCl), cryolite (Na$_3$AlF$_6$), fluorite (CaF$_2$), and atacamite. No molecules are present among the arrangement of the ions in halite, a naturally occurring form of sodium chloride. Each cation and anion is in octahedral coordination with its six closest neighbours. The NaCl structure is found in the crystals of many XZ-type halides, including sylvite (KCl) and chlorargyrite (AgCl). Some sulfides and oxides of XZ type crystallize in this structure type as well—for example, galena (PbS), alabandite (MnS), and periclase (MgO).

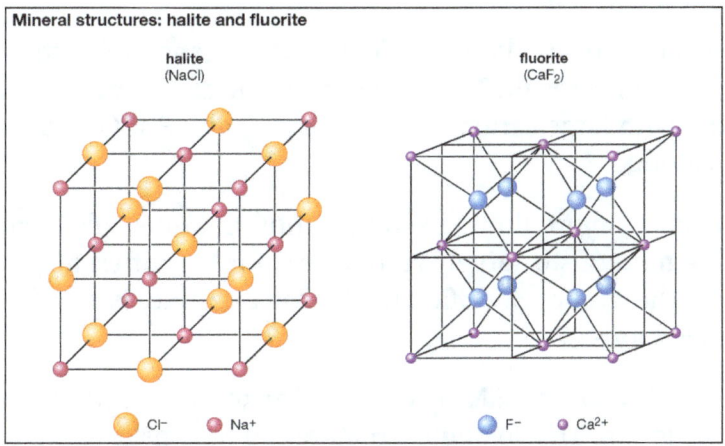

(A) The structure of halite, NaCl. (B) The structure of fluorite, CaF$_2$.

Several XZ$_2$ halides have the same structure as fluorite (CaF$_2$). In fluorite, calcium cations are positioned at the corners and face centres of cubic unit cells. (A unit cell is the smallest group of atoms, ions, or molecules from which the entire crystal structure can be generated by its repetition.) Each fluorine anion is in tetrahedral coordination with four calcium ions, while each calcium cation is in eightfold coordination with eight fluorine ions that form the corners of a cube around it. Uraninite (UO$_2$) and thorianite (ThO$_2$) are two examples of the several oxides that have a fluorite-type structure.

Carbonates

The carbonate minerals contain the anionic complex (CO$_3$)$^{2-}$, which is triangular in its coordination—i.e., with a carbon atom at the centre and an oxygen atom at each of the corners of an equilateral triangle. These anionic groups are strongly bonded individual units and do not share oxygen atoms with one another. The triangular carbonate groups are the basic building units of all carbonate minerals and are largely responsible for the properties particular to the class.

Carbonates are frequently identified using the effervescence test with acid. The reaction that results in the characteristic fizz, $2H^+ + CO_3^{2-} \rightarrow H_2O + CO_2$, makes use of the fact that the carbon-oxygen bonds of the CO$_3$ groups are not quite as strong as the corresponding carbon-oxygen bonds in carbon dioxide.

The common anhydrous (water-free) carbonates are divided into three groups that differ in structure type: calcite, aragonite, and dolomite. The copper carbonates azurite and malachite are the only notable hydrous varieties.

The members of the calcite group share a common structure type. It can be considered as a derivative of the NaCl structure in which the carbonate (CO$_3$) groups substitute for the chlorine ions and calcium cations replace the sodium cations. As a result of the triangular shape of the CO$_3$ groups, the structure is rhombohedral instead of isometric as in NaCl. The CO$_3$ groups are in planes perpendicular to the threefold c-axis, and the calcium ions occupy alternate planes and are bonded to six oxygen atoms of the CO_3 groups.

Members of the calcite group exhibit perfect rhombohedral cleavage. The composition $CaCO_3$ most commonly occurs in two different polymorphs: rhombohedral calcite with calcium surrounded by six closest oxygen atoms and orthorhombic aragonite with calcium surrounded by nine closest oxygen atoms.

When CO_3 groups are combined with large divalent cations (generally with ionic radii greater than 1.0 Å), orthorhombic structures result. This is known as the aragonite structure type. Members of this group include those with large cations: $BaCO_3$, $SrCO_3$, and $PbCO_3$. Each cation is surrounded by nine closest oxygen atoms.

The aragonite group displays more limited solid solution than the calcite group. The type of cation present in aragonite minerals is largely responsible for the differences in physical properties among the members of the group. Specific gravity, for example, is roughly proportional to the atomic weight of the metal ions.

Dolomite [$CaMg(CO_3)_2$], kutnohorite [$CaMn(CO_3)_2$], and ankerite [$CaFe(CO_3)_2$] are three isostructural members of the dolomite group. The dolomite structure can be considered as a calcite-type structure in which magnesium and calcium cations occupy the metal sites in alternate layers. The calcium (Ca^{2+}) and magnesium (Mg^{2+}) ions differ in size by 33 percent, and this produces cation ordering with the two cations occupying specific and separate levels in the structure. Dolomite has a calcium-to-magnesium ratio of approximately 1:1, which gives it a composition intermediate between $CaCO_3$ and $MgCO_3$.

Nitrates

The nitrates are characterized by their triangular $(NO_3)^-$ groups that resemble the $(CO_3)^{2-}$ groups of the carbonates, making the two mineral classes similar in structure. The nitrogen cation (N^{5+}) carries a high charge and is strongly polarizing like the carbon cation (C^{4+}) of the CO^3 group. A tightly knit triangular complex is created by the three nitrogen-oxygen bonds of the NO^3 group; these bonds are stronger than all others in the crystal. Because the nitrogen-oxygen bond has greater strength than the corresponding carbon-oxygen bond in carbonates, nitrates decompose less readily in the presence of acids.

Nitrate structures analogous to those of the calcite group result when NO_3 combines in a 1:1 ratio with monovalent cations whose radii can accommodate six closest oxygen neighbours. For example, nitratite ($NaNO_3$), also called soda nitre, and calcite exhibit the same structure, crystallography, and cleavage. The two minerals differ in that nitratite is softer and melts at a lower temperature owing to its lesser charge; also, sodium has a lower atomic weight than calcium, causing nitratite to have a lower specific gravity as well. Similarly, nitre (KNO_3), also known as saltpetre, is an analogue of aragonite. These are two examples of only seven known naturally occurring nitrates.

Borates

Minerals of the borate class contain boron-oxygen groups that can link together, in a phenomenon known as polymerization, to form chains, sheets, and isolated multiple groups. The silicon-oxygen (SiO_4) tetrahedrons of the silicates polymerize in a manner similar to the $(BO_3)^{3-}$ triangular groups of the borates. A single oxygen atom is shared between two boron cations (B^{3+}), thereby linking the BO_3 groups into extended units such as double triangles, triple rings, sheets, and chains. The oxygen atom is able to accommodate two boron atoms because the small boron cation has a bond strength to each oxygen that is exactly one-half the bond energy of the oxygen ion.

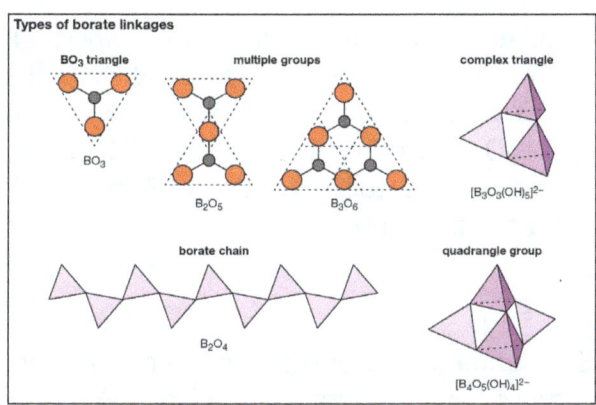

Various possible linkages of (A) BO_3 triangles to form (B,C) multiple groups and (D) chains in borates. Complex (E) triangle and (F) quadrangle groups are also shown. The group depicted in (F) occurs in borax.

Although boron is usually found in triangular coordination with three oxygens, it also occurs in fourfold coordination in tetrahedral groups. In addition, boron may exist as part of complex anionic groups such as $[B_3O_3(OH)_3]^{2-}$, consisting of one triangle and two tetrahedrons. Complex infinite chains of tetrahedrons and triangles are found in the structure of colemanite $[CaB_3O_4(OH)_3 \cdot H_2O]$; a complex ion composed of two tetrahedrons and two triangles, $[B_4O_5(OH)_4]^{2-}$, is present in borax $[Na_2B_4O_5(OH)_4 \cdot 8H_2O]$.

Sulfates

This class is composed of a large number of minerals, but relatively few are common. All contain anionic $(SO_4)^{2-}$ groups in their structures. These anionic complexes are formed through the tight bonding of a central S^{6+} ion to four neighbouring oxygen atoms in a tetrahedral arrangement around the sulfur. This closely knit group is incapable of sharing any of its oxygen atoms with other SO_4 groups; as such, the tetrahedrons occur as individual, unlinked groups in sulfate mineral structures.

Common sulfates	
Barite Group	
Barite	$BaSO_4$
Celestite	$SrSO_4$
Anglesite	$PbSO_4$
Anhydrite	$CaSO_4$
Gypsum	$CaSO_4 \cdot 2H_2O$

Members of the barite group constitute the most important and common anhydrous sulfates. They have orthorhombic symmetry with large divalent cations bonded to the sulfate ion. In barite ($BaSO_4$), each barium ion is surrounded by 12 closest oxygen ions belonging to seven distinct SO_4 groups. Anhydrite ($CaSO_4$) exhibits a structure very different from that of barite since the ionic radius of Ca^{2+} is considerably smaller than Ba^{2+}. Each calcium cation can only fit eight oxygen atoms around it from neighbouring SO_4 groups. Gypsum ($CaSO_4 \cdot 2H_2O$) is the most important and abundant hydrous sulfate.

Phosphates

Although this mineral class is large (with almost 700 known species), most of its members are quite rare. Apatite [$Ca_5(PO_4)_3(F, Cl, OH)$], however, is one of the most important and abundant phosphates. The members of this group are characterized by tetrahedral anionic $(PO4)^{3-}$ complexes, which are analogous to the $(SO_4)^{2-}$ groups of the sulfates. The phosphorus ion, with a valence of positive five, is only slightly larger than the sulfur ion, which carries a positive six charge. Arsenates and vanadates are similar to phosphates.

Silicates

The silicates, owing to their abundance on Earth, constitute the most important mineral class. Approximately 25 percent of all known minerals and 40 percent of the most common ones are silicates; the igneous rocks that make up more than 90 percent of Earth's crust are composed of virtually all silicates.

The fundamental unit in all silicate structures is the silicon-oxygen $(SiO_4)^{4-}$ tetrahedron. It is composed of a central silicon cation (Si^{4+}) bonded to four oxygen atoms that are located at the corners of a regular tetrahedron. The terrestrial crust is held together by the strong silicon-oxygen bonds of these tetrahedrons. Approximately 50 percent ionic and 50 percent covalent, the bonds develop from the attraction of oppositely charged ions as well as the sharing of their electrons.

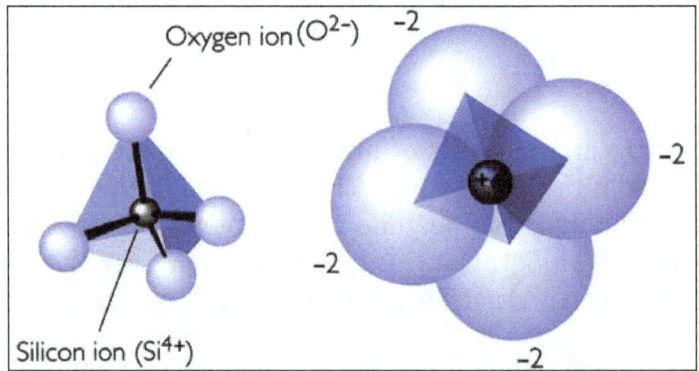

Two views of a closest-packed representation of the silicon-oxygen tetrahedron.

The positive charge (+4) of each silicon cation is satisfied by its four bonds to oxygen atoms. Each oxygen ion (O^{2-}), however, contributes only one-half of its total bonding energy to a silicon-oxygen bond, so it is capable of also bonding to the silicon cation of another tetrahedron. The SiO_4 tetrahedrons thereby become linked by shared oxygen atoms; this is referred to as polymerization. The degree and manner of polymerization are the bases for the variety present in silicate structures.

The silicates can be divided into groups according to structural configuration, which arises from the sharing of one, two, three, or all oxygen ions of a tetrahedron. Nesosilicates have isolated groups of SiO_4, while sorosilicates contain pairs of SiO_4 tetrahedrons linked into Si_2O_7 groups. Ring silicates, also known as cyclosilicates, are closed, ringlike silicates; the sixfold variety has composition Si_6O_{18}. Silicates that are composed of infinite chains of tetrahedrons are called inosilicates; single chains have a unit composition of SiO_3 or Si_2O_6, whereas double chains contain a silicon to oxygen ratio of 4:11. Phyllosilicates, or sheet silicates, are formed when three oxygen atoms are shared with adjoining tetrahedrons. The resulting infinite flat sheets have unit composition Si_2O_5. In structures where tetrahedrons share all their oxygen ions, an infinite three-dimensional network is created with an SiO_2 unit composition. Minerals of this type are called framework silicates or tectosilicates.

As a major constituent of Earth's crust, aluminum follows only oxygen and silicon in importance. The radius of aluminum, slightly larger than that of silicon, lies close to the upper bound for allowable fourfold coordination in crystals. As a result, aluminum can be surrounded with four oxygen atoms arranged tetrahedrally, but it can also occur in sixfold coordination with oxygen. The ability to maintain two roles within the silicate structure makes aluminum a unique constituent of these minerals. The tetrahedral AlO_4 groups are approximately equal in size to SiO_4 groups and therefore can become incorporated into the silicate polymerization scheme. Aluminum in sixfold coordination may form ionic bonds with the SiO_4 tetrahedrons. Thus, aluminum may occupy tetrahedral sites as a replacement for silicon and octahedral sites in solid solution with elements such as magnesium and ferrous iron.

Several ions may be present in silicate structures in octahedral coordination with oxygen: Mg^{2+}, Fe^{2+}, Fe^{3+}, Mn^{2+}, Al^{3+}, and Ti^{4+}. All cations have approximately the same dimensions and thus are found in equivalent atomic sites, even though their charges range from positive two to positive four. Solid solution involving ions of different charge is accomplished through coupled substitutions, thereby maintaining neutrality of the structures.

Nesosilicates

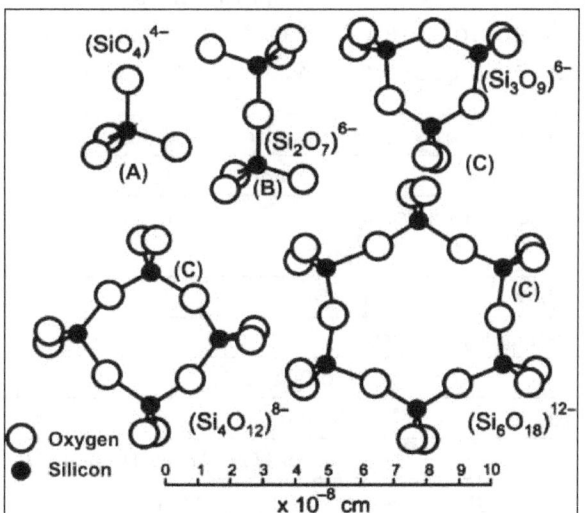

The silicon-oxygen tetrahedrons of the nesosilicates are not polymerized; they are linked to one another only by ionic bonds of the interstitial cations. As a result of the isolation of the tetrahedral groups, the crystal habits of these minerals are typically equidimensional so that prominent cleavage directions are not present. The size and charge of the interstitial cations largely determine the structural form of the nesosilicates. The relatively high specific gravity and hardness that are characteristic of this group arise from the dense packing of the atoms within the structure. Substitution of aluminum for silicon is normally quite low.

Sorosilicates

These minerals contain sets of two SiO_4 tetrahedrons joined by one shared apical oxygen. A silicon-to-oxygen ratio of 2:7 is consequently present in their structures. More than 70 minerals belong to the sorosilicate group, although most are rare. Only the members of the epidotegroup and vesuvianite are common. Both independent $(SiO_4)^{4-}$ and double $(Si_2O_7)^{6-}$ groups are incorporated into the epidote structure, as is reflected in its formula: $Ca_2(Al, Fe)Al_2O(SiO_4)(Si_2O_7)(OH)$.

Cyclosilicates

Silicon-oxygen tetrahedrons are linked into rings in cyclosilicate structures, which have an overall Si:O ratio of 1:3. There are three closed cyclic configurations with the following formulas: Si_3O_9, Si_4O_{12}, and Si_6O_{18}. The rare titanosilicate benitoite $(BaTiSi_3O_9)$ is the only mineral that is built with the simple Si_3O_9 ring. Axinite $[(Ca, Fe, Mn)_3Al_2(BO_3)(Si_4O_{12})(OH)]$ contains Si_4O_{12} rings, along with BO_3 triangles and OH groups. The two common and important cyclosilicates, beryl $(Be_3Al_2Si_6O_{18})$ and tourmaline (which has an extremely complex formula), are based on the Si_6O_{18} ring.

Inosilicates

This class is characterized by its one-dimensional chains and bands created by the linkage of SiO_4 tetrahedrons. Single chains may be formed by the sharing of two oxygen atoms from each tetrahedron, resulting in a structure with an Si:O ratio of 1:3. Two such chains that are aligned side by side with alternate tetrahedrons sharing an additional oxygen atom form bands of double chains. These structures have an Si:O ratio of 4:11. There are a number of silicate minerals, pyroxenoids, which have a similar Si:O ratio as pyroxene, but with structures that are not identical as the chains of silicon tetrahedra do not infinitely repeat. Two significant rock-forming mineral families display these structure types: the single-chain pyroxenes and the double-chain amphiboles.

Inosilicates: Common pyroxenes and amphiboles	
Pyroxenes	
Enstatite-orthoferrosilite series	
Enstatite	$MgSiO_3$
Orthoferrosilite	$FeSiO_3$
Diopside-hedenbergite series	
Diopside	$CaMgSi_2O_6$
Hedenbergite	$CaFeSi_2O_6$
Augite	$(Ca, Na) (Fe, Mg, Al) (Al, Si)_2O_6$
Sodium pyroxene group	
Jadeite	$NaAlSi_2O_6$
Acmite	$NaFe^{3+}Si_2O_6$
Amphiboles	
Anthophyllite	$(Mg, Fe)_7Si_8O_{22}(OH)_2$
Cummingtonite series	
Cummingtonite	$Fe_2Mg_5Si_8O_{22}(OH)_2$
Grunerite	$Fe_7Si_8O_{22}(OH)_2$

Tremolite series	
Tremolite	$Ca_2Mg_5Si_8O_{22}(OH)_2$
Actinolite	$Ca_2(Mg, Fe)_5Si_8O_{22}(OH)_2$
Hornblende	$(Ca, Na)_2(Mg, Fe, Al)_5(Si, Al)_8O_{22}(OH)_2$
Sodic amphibole group	
Glaucophane	$Na_2Mg_3Al_2Si_8O_{22}(OH)_2$
Riebeckite	$Na_2Fe_3^{2+}Fe_2^{3+}Si_8O_{22}(OH)_2$

The amphiboles and pyroxenes share the same cations and have many similar crystallographic, chemical, and physical properties: the colour, lustre, and hardness of analogous species are alike. A distinguishing factor between the two groups, the presence of the hydroxyl radical in the amphiboles, generally gives the double-chain members lower specific gravities and refractive indices than their single-chain analogues. Their crystal habits also are different: amphiboles exhibit needlelike or fibrous crystals, while pyroxenes take the form of stubby prisms. In addition, the different chain structures of the two groups result in different cleavage angles.

Pyroxenes occur in high-temperature igneous and metamorphic rocks. They crystallize at higher temperatures than their amphibole counterparts. A pyroxene formed early in the cooling of an igneous melt or in a metamorphic fluid may later combine with water at a lower temperature to form amphibole.

Phyllosilicates

These minerals display a two-dimensional framework of infinite sheets of SiO_4 tetrahedrons. An Si:O ratio of 2:5 results from the sharing of three oxygen atoms in each tetrahedron. Sixfold symmetry is exhibited in undistorted sheets. The silicate sheet framework is largely responsible for the following properties of the phyllosilicates: platy or flaky habit, single pronounced cleavage, low specific gravity, softness, and possible flexibility and elasticity of cleavage layers. Most minerals of this group contain hydroxyls positioned in the middle of the sixfold rings of tetrahedrons.

Many soil constituents, produced through rock weathering, possess a sheet structure. Phyllosilicate properties contribute greatly to the ability of soils to release and retain plant food, to reserve water from wet to dry seasons, and to accommodate organisms and atmospheric gases.

Tectosilicates

Almost 75 percent of Earth's crust is composed of minerals with the three-dimensional framework of the tectosilicates. All oxygen atoms of the SiO_4 tetrahedrons of members of this class are shared with nearby tetrahedrons, creating a strongly bound structure with an Si:O ratio of 1:2. Other than the zeolite group, which can accommodate water owing to the open nature of its structure, all members listed in the table are anhydrous.

Mineral Associations and Phase Equilibrium

Petrology, the scientific study of rocks, is concerned largely with identifying individual minerals in rocks, along with their abundance, grain size, and texture, because rocks typically consist of a variety of minerals. Such information is essential to an understanding of the history of any rock.

Petrological research requires a strong understanding of the principles of mineralogy and mineral identification and a thorough familiarity with the theoretical and experimental studies of rock origins. The present focuses on phase equilibrium, upon which the link between the study of minerals and the study of rocks is largely based.

A phase is a homogeneous substance that has a fixed composition and uniform chemical and physical properties. Only a mineral that displays no solid solution may therefore be considered a phase. Quartz (SiO_2), for example, is a low-temperature phase in the Si-O_2(SiO_2) system, and kyanite (Al_2SiO_5) is a high-pressure phase in the Al_2O_3-SiO_2(Al_2SiO_5) system. The term phase region is used when a mineral exhibits compositional variation, as in the solid-solution series between forsterite and fayalite. A phase may exist as a solid, liquid, or gas: H_2O, for example, occurs in the form of ice (solid), water (liquid), and steam (gas).

Equilibrium refers to the stable coexistence of two or more phases and is established relative to time. If two phases in a mixture of water and ice coexist so that the amount of each is fixed indefinitely, they are said to be in equilibrium. The minerals of some rocks have existed together since their formation for periods of several million years, yet one cannot always ascertain if these rock constituents are in equilibrium or are still undergoing changes.

A determining factor of the equilibrium state of minerals is the presence (or absence) of a reaction rim, which is a region separating two or more minerals and consisting of the products of a reaction between them. The absence of any observable reaction rims between minerals that physically touch each other suggests that they were in equilibrium at the time when the rock formed. Additional chemical data regarding elemental distribution between the minerals is necessary to verify this assumption. In contrast, the presence of megascopically or microscopically visible rims indicates that some minerals were not in equilibrium. Garnet, for example, may react with coexisting biotite to produce a chlorite rim between them, revealing that the two minerals were not always in equilibrium. An experimental petrologist must assign some period of time after which the absence of further changes between phases will indicate that equilibrium has been reached. The time period is variable, depending on the speed of the reactions involved and in part on the patience of the investigator; it may range from a few hours to several years.

Reaction Rim

Components are the minimum number of independent chemical species that are necessary to describe the compositions of all the phases present in a system. The compound H_2O is generally used as the sole component defining the H_2O system, although H_2 and O_2 define the chemical system as well. In examinations of the stability fields of $MgSiO_3$ (enstatite), $MgSiO_3$ is normally used as the component rather than the three elements, Mg, Si, and O, or the two oxides, MgO and SiO_2. The three components generally used in the pyroxene system CaO-MgO-FeO-SiO_2 are $CaSiO_3$-$MgSiO_3$-$FeSiO_3$.

Assemblage and the Phase Rule

In the early stages of the study of a rock, the constituent minerals of the rock must be identified. Orthoclase, albite, quartz, and biotite may be found in an igneous granite. By examining the granite's texture, one may conclude that the four minerals crystallized at approximately the same elevated temperature and that orthoclase-albite-quartz-biotite is its mineral assemblage. The term assemblage is frequently applied to all minerals included in a rock but more appropriately should be used for those minerals that are in equilibrium (and are known more specifically as the equilibrium assemblage). The granite discussed above may display surficial cavities that are lined by several clay minerals and limonite (a hydrous ironoxide). The original high-temperature granite was altered to form the low-temperature clay minerals and limonite; there are consequently two distinct assemblages present in the rock: the high-temperature orthoclase-albite-quartz-biotite assemblage and the low-temperature assemblage of clay minerals and limonite.

Metamorphic rocks also may contain separate assemblages. A shale that at low temperatures was composed of a sericite-kaolin-dolomite-quartz-feldspar assemblage can become metamorphosed at higher temperatures to produce a garnet-sillimanite-biotite-feldspar assemblage.

An assemblage thus consists of minerals that formed under the same or quite comparable conditions of pressure and temperature. In practice, minerals that physically touch one another with no reaction rims or alteration products are included in the assemblage. It is likely that the minerals satisfying these conditions are in equilibrium, but additional chemical tests are commonly necessary to define the equilibrium assemblage without ambiguity.

Phase systems are governed by a phase rule, which defines the number of minerals that may coexist in equilibrium: $F = C - P + 2$, where F is the variance, or number of degrees of freedom, C is the number of independent components, and P is the number of phases. Applying this rule to a three-phase, three-component system, F is 2. This indicates that two parameters—e.g., pressure and temperature—may be varied independently of one another without altering the number of phases.

Phase Diagrams

Phase (or stability) diagrams are used to illustrate the conditions under which certain minerals are stable. They are graphs that show the limiting conditions for solid, liquid, and gaseous phases of a single substance or of a mixture of substances while undergoing changes in pressure and temperature or in some other combination of variables. The following are examples of phase diagrams employed in the study of igneous, metamorphic, and sedimentary rocks.

Use in Igneous Petrology

In the field of igneous petrology, the researcher commonly employs a phase equilibriumapproach to compare the mineral assemblages found in naturally occurring and syntheticrocks. Much can be learned from studying the melting of an igneous rock and the reverse process, the crystallization of minerals from a melt (liquid phase). Graphic representations of systems with a liquid phase are called liquidus diagrams. The dashed contours of a liquidus diagram, called isotherms, represent temperatures at which a mineral melts. They define what is known as a liquidus surface. As temperatures decrease, the minerals will crystallize in the manner defined by the arrows on the boundaries separating the different mineral phases. A careful study of the crystalline products formed upon the cooling of melts of specific compositions allows the igneous petrologist to compare such results with minerals observed in natural igneous rocks.

Use in Metamorphic Petrology

Pressure-temperature (P-T) phase diagrams are applied in the study of the conditions under which metamorphic rocks originate. They illustrate the equilibrium relationships among various mineral phases in terms of pressure and temperature. The minerals that are separated by a reaction curve may exist in equilibrium at the conditions occurring along lines present within the diagram. For example, the reaction curves for Al_2SiO_5 and for muscovite + quartz $\longleftrightarrow$ potassium-feldspar + sillimanite + H_2O are significant in metamorphic rocks that have a high aluminum oxide (Al_2O_3) content as compared to other components (e.g., calcium oxide [CaO], magnesium oxide [MgO], and ferrous oxide [FeO]). Shales enriched in clay minerals contain a rather large amount of aluminum oxide, and during metamorphism of the shalemineral reactions and recrystallization occur. In their metamorphic form, shales appear as pelitic schists, and these may include significant amounts of sillimanite, muscovite, and quartz. Such a schist may have equilibrated under a certain set of pressure and temperature conditions.

Theoretical calculations are combined with experimental observations to arrive at phase diagrams. In laboratory experiments conducted on the three polymorphs of Al_2SiO_5, chemicals of high purity are most often used as the starting materials, but extremely pure minerals may be substituted. A specimen of gem-grade kyanite that does not contain any inclusions may be reacted at high temperatures to form sillimanite. The placement of the reaction curve between the kyanite and sillimanite fields is determined by the first instance of sillimanite formation from kyanite and also the initial stage of the reverse reaction. X-ray powder diffraction and optical microscopic techniques are employed to estimate the conditions under which these reactions commence, but the experimental methods are subject to some degree of uncertainty. Therefore, the reaction curves, although commonly drawn as narrow lines, may actually represent wider reaction zones. Also, the naturally occurring system is more complex than the simplified version devised in the laboratory, and so difficulty arises when attempting to relate the two. A P-T diagram serves mainly as a tool in evaluating the conditions of metamorphism, such as pressure and temperature.

Use in Sedimentary Petrology

Phase diagrams can also be helpful in the assessment of physical and chemical conditions that prevailed during the deposition of a chemical sedimentary sequence. Atmospheric conditions are characterized by low temperatures and pressures, and under such conditions stability fields of minerals can often conveniently be expressed in terms of Eh (oxidation potential) and pH (the negative logarithm of the hydrogen ion concentration [H^+]; a pH of 0–7 indicates acidity, a pH of 7–14 indicates basicity, and neutral solutions have a pH of 7).

Stability relations for some iron oxides and iron sulfides are often presented at atmospheric conditions, 25 °C (77 °F) and one standard atmosphere pressure. (One standard atmosphere of pressure equals

760 millimetres, or 29.92 inches, of mercury.) A high Eh value corresponds to a compound stable under oxidizing conditions, such as hematite, while a low Eh value indicates a mineral that occurs in reducing environments, such as magnetite. Pyrite and pyrrhotite, two sulfide minerals, occur at low Eh values and at pH values of 4–9. Lines separating the fields of an Eh-pH diagram represent conditions under which the two minerals may exist in equilibrium. Hematite and magnetite, for example, are often found together in iron-bearing sediments. Eh-pH diagrams are valuable in providing information regarding the chemical and physical environments that existed during atmospheric weathering and during chemical sedimentation and diagenesis of sediments deposited by water at temperatures of 25 to about 100 °C (77 to about 212 °F) and at a pressure of approximately one atmosphere. The coexistence of hematite and magnetite common in Precambrian iron-bearing rocks (those formed from 4.6 billion to 541 million years ago) may enable investigators to estimate variables such as Eh and pH that prevailed in the original ancient sedimentary basin.

Mineraloid

A mineraloid is a naturally occurring, inorganic solid that does not exhibit crystallinity. It may have the outward appearance of a mineral, but it does not have the "ordered atomic structure" required to meet the definition of a mineral. Some mineraloids also lack the "definite chemical composition" required to be a mineral.

To be considered a mineral, a material must meet the following five requirements:

1. Naturally occurring;

2. Inorganic;

3. Solid;

4. Ordered atomic structure;

5. Definite chemical composition (can vary within a limited range).

Minerals are "crystalline." In other words, they have an ordered atomic structure. In contrast, mineraloids are "amorphous." This means that their internal atomic structure is not ordered.

Without the ordered atomic structure, mineraloids never produce well-formed crystals. They also do not exhibit the property of cleavage because they lack internal planes of weakness.

Common opal is a mineraloid. It is an amorphous silica with a chemical composition of $SiO_2 \cdot nH_2O$. It has a conchoidal fracture that is characteristic of an amorphous glass.

Examples of Mineraloids

There are a number of familiar materials that can be classified as mineraloids. For example, opal is an amorphous hydrated silica with a chemical composition of $SiO_2 \cdot nH_2O$. The "n" in its formula indicates that the amount of water is variable. Therefore, opal is a mineraloid.

Obsidian and pumice are igneous rocks that solidified so rapidly from a melt that their atoms were unable to move into an ordered atomic structure. Instead, they rapidly formed a random network of atoms known as a "glass." Obsidian and pumice are amorphous, and their compositions can vary dramatically from one location to another and from one volcanic eruption to the next. Obsidian and pumice are also mineraloids.

Obsidian is a mineraloid. It is a volcanic glass that cools so rapidly that atoms do not have time to arrange themselves into a crystalline solid. Instead, they form an amorphous, randomly bonded network.

Pumice is an expanded volcanic glass. It forms during explosive eruptions of gas-charged magma. It is ejected from the volcano in a sudden blast and cools so quickly that bubbles of gas are trapped within the amorphous glass.

Mineraloids from the Sky

Tektites and moldavites are varieties of natural glass that formed from the impact of an asteroid or comet. These objects struck the Earth at hypervelocity, and the force of their impact produced a tremendous amount of heat energy. The explosion that occurred upon impact flash-melted the target rock and produced a shower of molten material over thousands to millions of square miles. The molten material's temperature dropped quickly as it flew through the air - so quickly that the melts solidified without forming crystals.

Libyan desert glass is a similar material thought to be caused by an impact in a sandy area. Fulgurite and the associated material known as lechatelierite are produced when lightning strikes the Earth in a sandy environment. These strikes instantly melt the sand, which then rapidly solidifies as amorphous silica. These materials are rapidly cooled glassy mineraloids.

Tektites are pieces of black glass formed by an impact somewhere between Australia and Southeast Asia about 800,000 years ago. The specimen in the above image is a tektite from the Australasian strewnfield. Millions of tektites, ranging from sand-size grains to fist-size nodules, have been found in that area. Their surfaces are often marked with the same surface regmaglypts seen on iron meteorites.

Moldavite is another type of impact glass, which was formed about 15 million years ago when an asteroid struck the area that is now eastern Europe. The green glass is now found and valued by collectors. Transparent pieces with good clarity are sometimes cut as gems.

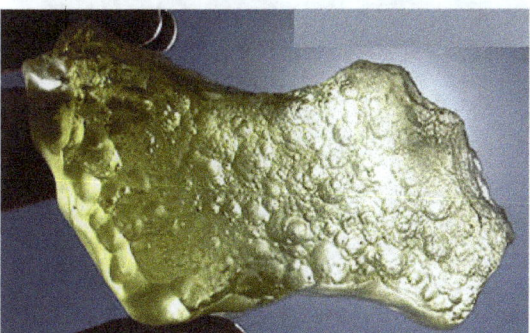

Libyan Desert Glass is a yellow glass found scattered over the desert near the border between Egypt and Libya. It is believed to have formed in the seconds after an asteroid impact about 29 million years ago. Large amounts of desert surface were flash-melted by the heat of the impact and scattered over the surrounding land.

Mineraloid-forming Environments

Most mineraloids form at the low temperatures and low pressures found at Earth's surface and in shallow subsurface environments. Materials such as opal, psilomelane, chrysocolla, limonite, and a wide variety of supergene materials crystallize from gels or colloids in the shallow sub-surface. Many of these materials will eventually transform into minerals with time, heat, or pressure. These low-temperature mineraloids often have a mammillary (smoothly rounded or hemispherical), botryoidal (grape-like clusters), pisolitic (pea-like clusters), or stalactitic (icicle-like) habit.

Psilomelane is a black hydrous manganese oxide that often contains barium and potassium. It is an amorphous material and an ore of manganese.

Limonite is an amorphous mineraloid. It is a hydrated iron oxide.

Mercury is a liquid at room temperature, but when cooled to -38.8 °C,
it crystallizes into a solid. Solid mercury meets all of the requirements of a mineral,
and thus some people consider liquid mercury to be a mineraloid.

Water and mercury are often classified as mineraloids. They are the only two natural inorganic substances that have a definite chemical composition and are liquids at room temperature. They are also the only two liquids that crystallize into minerals within the range of temperatures and pressures encountered at Earth's surface. Water crystallizes into the mineral "water ice" when cooled to 0 °C. Mercury crystallizes into solid mercury at -38.8 °C. Due to the fact that they crystallize into minerals, some mineralogists include water and mercury in the mineraloids group.

Water is considered to be a mineraloid by many mineralogists.
It crystallizes into water ice when cooled to 0 °C.

Radiolarite is a sedimentary rock that forms from the accumulation of microscopic radiolarian tests and is thus of organic origin. This specimen is from the Windalia Radiolarite, a rock unit that formed on a marine shelf in an area that is now part of the Kennedy Ranges of Western Australia. When hard and tough it can be cut into cabochons, beads, tumbled stones, and other ornamental objects. It is sold as a gem material under the trade name "mookaite." Although it is a variety of chalcedony, some consider it a mineraloid because of its organic origin.

Organic Mineraloids

If you read information about mineraloids written by a variety of authors, you will discover that some authors include organic materials, such as amber and jet, in their list of mineraloids. Some mineralogists agree with such classifications, but others feel this stretches the definition of a mineraloid too far.

Amber is a fossil plant resin found in sediments and sedimentary rocks in many parts of the world. It is hard, brittle, translucent to transparent, and is often cut as a gemstone. It has the appearance of a mineral, but lacks an ordered internal structure and lacks a definite chemical composition. Furthermore, it is organic. It fails three of the five tests for being a mineral. Should it be called a "mineraloid"?

Jet is a rare type of dark black coal. It has a smooth texture that accepts a bright polish, which is why it is often cut as a gemstone. It has the outward appearance of a mineral but lacks a crystalline structure and a definite chemical composition. It is also organic. Should it be called a "mineraloid"?

A number of very tiny organisms, such as diatoms and radiolarians, produce a thin shell of amorphous silica known as a "test." When these organisms die, their tests sink to the bottom. When the tests are the dominant material that accumulates, the sediment is known as "ooze." If buried and lithified, the ooze can transform into rocks such as diatomite and radiolarite.

Properties of Minerals

Chemical Properties

Every mineral contains a defined ratio of specific molecules in its structure. For example, a water molecule is composed of H_2O - two hydrogen atoms and one oxygen atom. When water molecules are grouped together, they form a body of water. Trace amounts of foreign molecules not inherent to a mineral's makeup are known as impurities. Impurities can slightly alter physical properties such as color.

Atoms join together based on their positive and negative charges. This is caused by the amounts of protons (positive charges) or electrons (negative charges) they contain.

Chemical Formula

Every mineral has a unique arrangement of elements within its inherent structure. This arrangement of atoms determines a mineral type. All minerals have a chemical formula, which is an analysis of the types and amounts of elements present in a mineral. Every element has a one or two letter abbreviated term. For example oxygen is "O", and gold is "Au".

The chemical formula of the mineral Hematite is Fe_2O_3. The letters describe the element type (Fe = iron, O = oxygen), and the subscripted numbers describe the amount of those atoms in each molecule. A Hematite molecule has 2 iron (Fe) atoms and 3 oxygen (O) atoms. If there is no number written after an element, it means there is only one atom of that element present.

Radicals

Radicals, or polyatomic ions, are special types of compounds. They act as if they were a single element when they join other elements to form molecules. There are many radicals, some of the most familiar are given below.

Carbonate radical	CO_3
Sulfate Radical	SO_4
Chromate Radical	CrO_4
Hydroxyl Radical	OH

Radicals are treated like any element when written in a chemical formula. Radicals are sometimes surrounded with a parenthesis when written in a chemical formula, and the number after the parenthesis describes how many of these radicals are present within each mineral's molecular breakdown. For example, the mineral Talc has a chemical formula of $Mg_3Si_4O_{10}(OH)_2$. The subscripted

2 at the end of the (OH) describes that there are two hydroxyl radicals within each molecule of Talc. If there is no subscripted number written after a radical, that means there is only one radical present in that mineral's molecule.

Hydrous Minerals

Minerals containing water in their structure are known as hydrous minerals. The hydrous mineral Gypsum has a chemical formula of "$CaSO_4 \cdot 2H_2O$". The large number 2 in front of the H_2O signifies that there are two water (H_2O) molecules for every molecule of $CaSO_4$. The dot in between $CaSO_4$ and $2H_2O$ indicates that these are two separate molecules, but they are rationally proportionate.

The letter "n" is used to describe a variable amount of water in the structure of a mineral. For example, the hydrous mineral Opal has a loosely defined composition with an inconsistent amount of water in its structure, thus its chemical formula is written as "$SiO_2 \cdot nH_2O$".

Some minerals, such as Torbernite, have a varying amount of water within fixed limits. The chemical formula for Torbernite is "$Cu(UO_2)(PO4)_2 \cdot 8\text{-}12H_2O$". The $8\text{-}12H_2O$ indicates that there can be 8 to 12 water molecules for every $Cu(UO_2)(PO_4)_2$ molecule.

Mineral Series

A number of minerals contain a varying amount of two or more elements. For example, the mineral Aurichalcite, which has a chemical formula of $(Zn,Cu)_5(CO_3)_2(OH)_6$ contains an unspecific varying amount of zinc (Zn) and copper (Cu). This is indicated by comma separating the Zn from the Cu. If a chemical formula with two elements in parenthesis is separated by a comma, the number of those elements vary. Aurichalcite has a variable amount of zinc of copper where the combination of both these elements totals five. The more dominant element is usually listed first.

Often, when the elements in a mineral vary, a series is formed. A series consists of a group of mineral in which one of the elements varies. For example, the Spinel series contains four members, with a series formula of $(Mg,Zn,Fe,Mn)Al_2O_4$. There are four end members of this series. Intermediary forms that are a combination of two or more also exist.

Spinel End Members		
Spinel	$MgAl_2O_4$	Magnesium Spinel
Gahnite	$ZnAl_2O_4$	Zinc Spinel
Hercynite	$FeAl_2O_4$	Iron Spinel
Galaxite	$MnAl_2O_4$	Manganese Spinel

Some intermediary members of the spinel group have designated names, such as Gahnospinel $(Mg,Zn)Al_2O_4$, which is a mixture of Spinel ($MgAl_2O_4$) and Gahnite ($ZnAl_2O_4$); while other mixtures lack a designated name and are just called under the umbrella of Spinel. Intermediary members are sometimes scientifically recognized as individual mineral species by the IMA, though they are more often not recognized (Gahnospinel is not recognized as a scientifically distinct mineral by the IMA).

Many mineral series form solid solutions. In a solid solution, there are intermediate members between the two end members. Individual intermediary members may be given names, while others may not. An example is the Olivine group, with a chemical formula of $(Mg,Fe)_2SiO_4$. It contains Forsterite, (Mg_2SiO_4), and Fayalite, (Fe_2SiO_4) as the two end members. Very rarely are the Olivine members pure; most are somewhere along the intermediary scale within the solid solution series, with a varying percentage of Mg and Fe.

Atomical Variations

There are certain elements that come in slight modifications. Some elements have different amounts of electrons in atoms of the same element. Some minerals, such as Babingtonite, contain two types of the same element. The chemical formula for Babingtonite is written $Ca_2Fe^{2+}Fe^{3+}SiO_{14}(OH)$. The superscript form of $^{2+}$ and $^{3+}$ next to the iron (Fe) distinguishes the different types of iron. $^{2+}$ means the iron has 2 more electrons than protons, and $^{3+}$ means it has three more electrons. The amount of electrons affects the chemical bonding of any element with variable electrons. Anytime there is more than one variable atom in a molecule, a superscript number representing the type of atom (i.e. how many electrons over protons) is written after the element symbol.

Chemical Formula Variables

The chemical formula of some minerals may be written in different formats and styles. It is not uncommon to see chemical formulas written differently for the same mineral. For example, the chemical formula for the mineral Dioptase is usually written $CuSiO_2(OH)_2$. However, its formula may also be written H_2CuSiO_4. Both formulas mean the exact same thing, as there is an equal amount of all the atoms in both formulas.

Sometimes a formula may be reduced or expanded by using multiplication or division. (A reduced formula is known as an empirical formula in chemistry.) For example, the chemical formula of Sodalite is commonly written as $Na_4Al_3Si_3O_{12}Cl$. The silicon and oxygen form a radical, so they are bound together as a single atom. Since the amount of silicon and oxygen is divisible by three $(Si_{3\div3}O_{12\div3} = SiO_4)$, the chemical formula can also be written as $Na_4Al_3(SiO_4)_3Cl$.

Variable Formulas

Some minerals may contain an element that partially replaces an inherent element. For example, the mineral Adamite, $Zn_2(AsO_4)(OH)$, often contains small amounts of Cu (copper) and Co (cobalt) replacing some Zn (zinc). These elements are not mentioned in the chemical formula, as they do not compromise a significant portion and are only occasionally present. Therefore, an additional formula, known as the variable formula has been developed for this guide to state the occasional presence of these elements.

The variable formula displays the regular formula with the additional elements that are occasionally present. The inherent element is listed first in the parenthesis, and the occasional replacement elements are listed after and underlined. For example, the variable formula of Adamite is $(Zn,Cu,Co)_2(AsO_4)(OH)$. The Zn (zinc) is always present, and the Cu and Co may be present in small amounts or may not be.

Some minerals without a variable formula may still have variable elements, but those elements are too rare or insignificant to be reckoned. For example, the mineral Fluorite occasionally has contains traces of Ce (cerium) and Y (yttrium), but there is no variable formula since these combinations are too uncommon.

Physical Properties

The physical characteristics of minerals include traits which are used to identify and describe mineral species. These traits include color, streak, luster, density, hardness, cleavage, fracture, tenacity, and crystal habit.

Certain wavelengths of light are reflected by the atoms of a mineral's crystal lattice while others are absorbed. Those wavelengths of light which are reflected are perceived by the viewer to possess the property of color. Some minerals derive their color from the presence of a particular element within the crystal lattice. The presence of such an element can determine which wavelengths of light are reflected and which are absorbed. This type of coloration in minerals is termed idiochromatism; different samples of an idiochromatic mineral species will all display the same color. Other minerals are colored by the presence of certain elements in mixture. Different samples of such a species may exhibit a range of similar colors. Still other mineral species may usually be colorless, but may display several different and startling colors when trace amounts of impurities, or elements which are not an integral part of the crystalline lattice, are present. Coloration which is caused by the presence of an element foreign to the crystal lattice, whether in mixture or in trace amounts, is termed allochromatism. Certain elements are strong pigmenting agents and may lend vivid colors to specimens when they are present, whether as a part of the crystal lattice, in mixture, or as an impurity. These elements are termed the chromophores.

Streak is the color which a mineral displays when it has been ground to a fine powder. Trace amounts of impurities do not tend to affect the streak of a mineral, so this characteristic is usually more predictable than color. Two different specimens of the same species may be expected to possess the same streak, whereas they may display different colors.

Minerals are either opaque or transparent. A thin section of an opaque mineral such as a metal will not transmit light, whereas a thin section of a transparent mineral will. Typically those minerals which possess metallic bonding are opaque whereas those where ionic bonding is prevalent are transparent. Relative differences in opacity and transparency are described as luster. The characteristic of luster provides a qualitative measure of the amount and quality of light which is reflected from a mineral's exterior surfaces. Luster thus describes how much the mineral surface 'sparkles'.

The property of density is defined as mass per unit volume. Certain trends exist with respect to density which may sometimes aid in mineral identification. Native elements are relatively dense. Minerals whose chemical composition contains heavy metals, or atoms possessing an atomic number greater than iron (Fe, atomic number 26), are relatively dense. Species which form at high pressures deep within the earth's crust are in general more dense than minerals which form at lower pressures and shallower depths. Dark-colored minerals are typically fairly dense whereas light-colored ones tend to be less dense.

Hardness is defined as the level of difficulty with which a smooth surface of a mineral specimen may be scratched. Hardness has historically been measured according to the Mohs scale. Mohs' method relies upon a scratch test to relate the hardness of a mineral specimen to the hardness of one of a set of reference minerals. Hardness may also be measured according to the more quantitative but less accessible diamond indentation method.

Cleavage refers to the splitting of a crystal along a smooth plane. A cleavage plane is a plane of structural weakness along which a mineral is likely to split. The quality of a mineral's cleavage refers both to the ease with which the mineral cleaves and to the character of the exposed surface. Not every mineral exhibits cleavage.

Fracture takes place when a mineral sample is split in a direction which does not serve as a plane of perfect or distinct cleavage. A mineral fractures when it is broken or crushed. Fracture does not result in the emergence of clearly demarcated planar surfaces; minerals may fracture in any possible direction.

The characteristic of tenacity describes the physical behavior of a mineral under stress or deformation. Most minerals are brittle; metals, in contrast, are malleable, ductile, and sectile.

The term crystal habit describes the favored growth pattern of the crystals of a mineral species. The crystals of particular mineral species sometimes form very distinctive, characteristic shapes. Crystal habit is also greatly determined by the environmental conditions under which a crystal develops.

Color

When different wavelengths of visible light are incident upon the eye they are perceived as being of different colors. Three different varieties of color receptors in the eye correspond to light possessing wavelengths of approximately 660 nm (red), 500 nm (green), and 420 nm (blue-violet). The eye then interprets the color of incident light according to which color receptors have been stimulated. For example, if monochromatic light which stimulated the red and green color receptors equally and did not affect the blue-violet receptors was detected, then the eye would interpret this light as possessing a wavelength halfway between those of red and green light. The eye would therefore register an incident light wave with a wavelength of approximately 580 nm and the viewer would percieve the incoming light as yellow. Incident polychromatic light which stimulated the red and green color receptors equally and did not affect the blue-violet ones would also be interpreted as yellow light, regardless whether or not the incoming light actually contained a component with a wavelength close to 580 nm. The incident polychromatic light might possess only a red and a green component of equal intensity; it would nevertheless be interpreted by the eye as yellow light. The phenomenon called color is thus a description of the differentiation by the eye between various wavelengths and combinations of wavelengths of visible light.

When light is incident upon a mineral specimen, some wavelengths are absorbed by the atoms of the crystal lattice while others are reflected. Those wavelengths which were not absorbed are reflected off of the mineral's surfaces and enter the eye of the viewer. The color which is perceived by the viewer depends on the wavelengths of light which are reflected rather than absorbed by the mineral. The property of color in minerals is thus due to the absorption of particular wavelengths of light and the reflection of others by the atoms of the crystal lattice.

The color exhibited by certain mineral species may depend upon which crystallographic axis is transmitting the light. Such species may demonstrate several different colors as light is transmitted along various different axes. This phenomena of directionally selective absorption is termed pleochroism.

Idiochromatism and the Chromophores

The color of many mineral species is derived directly from the presence of one or more of the elements which constitute the crystal lattice. The color of such minerals is a fundamental property directly related to the chemical composition of the species. Minerals which exhibit this type of coloration are called idiochromatic minerals. Idiochromatic coloration is a property possessed by a mineral species as a whole. In such species color can successfully be utilized as a means of identification.

Ions of certain elements are highly absorptive of selected wavelengths of light. Such elements are called chromophores; they possess strong pigmenting capabilities. The elements vanadium (V), chromium (Cr), manganese (Mn), iron (Fe), cobalt (Co), nickel (Ni), and copper (Cu) are chromophores. A mineral whose chemical formula stipulates the presence of one or more of these elements may possess a vivid and distinctive color.

Examples of idiochromatic minerals abound. For instance, the copper carbonate malachite is consistently green; the copper carbonate azurite and the copper silicate chrysocolla are each a distinctive and predictable blue. Rhodochrosite is always red or pink; samples of sulphur are a bright, recognizable yellow. Each of these distinctive colors is due to the fact that the chemical composition which defines the mineral species specifies inclusion of one of the chromophores within the lattice structure.

Allochromatism

Most minerals which are composed entirely of elements other than the chromophores are nearly colorless. However, certain specimens are sometimes observed to possess vivid coloration. Color in such instances is due to the presence of an impurity. If one of the chromophores is present within a mineral whose chemical formula does not include it, then the foreign element constitutes an impurity or a defect in the lattice structure. Coloration in minerals which is due to the presence of a foreign element is termed allochromatism. In such cases the color of the mineral may differ radically from the nearly colorless shade expected of the species.

Some minerals demonstrate a range of colors due to the presence in mixture of one of the chromophores. For example, the substitution of a quantity of iron for zinc atoms within the crystal lattice of sphalerite (ZnS) implements a change from white to yellow in the color of the mineral. Proportionally larger inclusions of iron will progressively result in a brown and eventually a black mineral specimen. In such cases the color of the sample is directly proportional to the amount of the pigmenting element which is present in the crystal lattice.

Not all allochromatism in minerals is due to presence of substantial amounts of a chromophore in mixture, however. The property of color may sometimes be highly dependent on the inclusion of trace amounts of impurities. The presence of even a minute quantity of a chromophore within the crystal lattice can cause a mineral specimen to exhibit vivid color. For example, trace inclusions of chromium (Cr) in beryl are responsible for the deep green of emerald, while the purple of amethyst is due to trace amounts of iron (Fe) in quartz and the pink of rose quartz is due to trace inclusions of titanium (Ti). Samples of the mineral corundum which include tiny amounts of chromium are deep red, and the gem is then called a ruby, while samples containing iron or titanium impurities produce blue gems termed sapphire.

Trace amounts of an impurity do not affect the basic chemical composition or the chemical formula of a mineral, and thus do not affect its classification as a species. Trace amounts of the various chromophores, however, can cause several samples of a single species to differ radically in color. (Beryl, corundum, and quartz provide examples of this possibility.) Because it varies so widely, color is a property which is sometimes of little use in identification. However, the idiochromatic minerals are consistently of distinctive color. The green of malachite, the blue of azurite, the pink of rhodocrosite, and the yellow of sulphur are easily recognized and are therefore quite useful in the identification of these species.

Streak

Streak is the color of a mineral substance when it has been ground to a fine powder. Typically an edge of the sample will be rubbed across a porcelain plate, leaving behind a 'streak' of finely ground material. The material in a streak sample thus consists of a powder composed of randomly oriented microscopic crystals rather than a lattice structure containing the uniformly oriented unit cells which compose a macroscopic crystal.

Although color is a property which may vary widely between two different specimens of the same mineral, streak generally varies little from sample to sample. The presence of trace amounts of an impurity may radically affect the property of color in a macroscopic crystal because each unit cell is aligned within the crystal structure, thereby forming a diffraction grating. Minute amounts of a strongly absorptive impurity within the structure may highly affect which wavelengths of light are reflected from this diffraction grating. This change may greatly modify the absorption of certain wavelengths of incoming light, altering the percieved color of the specimen. In a streak sample, however, each of the microscopic crystal grains of the sample is randomly oriented and the presence of an impurity does not greatly affect the absorption of incoming light. Because it is not typically affected by the presence of an impurity, streak is a more reliable identification property than is color.

Luster

Minerals may be categorized according to whether they are opaque or transparent. A thin section of an opaque mineral such as a metal will not transmit light, whereas a thin section of a transparent mineral will. The absorption index of an opaque mineral is high. Light which is incident upon an opaque mineral such as a metal is unable to propagate through the mineral due to this high rate of absorption, and will thus be reflected. Opaque minerals typically reflect between 20% to 50% or more of the light incident upon them. In contrast, most of the light which is incident upon a transparent mineral passes into and through the mineral; transparent minerals may reflect as little as 5% of the incident light and as much as 20%. Typically those minerals which possess metallic bonding are opaque whereas those where ionic bonding is prevalent are transparent.

Relative differences in opacity and transparency are described as luster. The term luster refers to the quantity and quality of the light which is reflected from a mineral's exterior surfaces. Luster provides an assessment of how much the mineral surface 'sparkles'. This quality is determined by the type of atomic bonds present within the substance. It is related to the indices of absorption and refraction of the material and the amount of dispersion from the crystal lattice, as well as the texture of the exposed mineral surface.

Minerals are primarily divided into the two categories of metallic and nonmetallic luster. Minerals possessing metallic luster are opaque and very reflective, possessing a high absorptive index. This type of luster indicates the presence of metallic bonding within the crystal lattice of the material. Examples of minerals which exhibit metallic luster are native copper, gold, and silver, galena, pyrite, and chalcopyrite. The luster of a mineral which does not quite possess a metallic luster is termed submetallic; hematite provides an example of submetallic luster.

The property of streak can aid in distinguishing whether a specimen has a metallic or a nonmetallic luster. Metals tend to be soft, implying that more powdered material may be obtained from the streak sample of a metal than a nonmetal. Metals are also opaque, transmitting no light. Minerals which possess a metallic luster therefore tend to exhibit a thick, dense, dark streak whereas those which possess a nonmetallic luster tend to produce a thinner, less dense streak which is also lighter in color.

Adjectives such as "vitreous', 'dull', 'pearly', 'greasy', 'silky' or 'adamantine' are frequently used to describe various types of nonmetallic luster.

Dull or Earthy

Minerals of dull or earthy luster reflect light very poorly and do not shine. This type of luster is often seen in minerals which are composed of an aggregate of tiny grains.

Resinous

A surface of resinous luster possesses a sheen resembling that of resin. Such materials have a refractive index greater than 2.0. Sphalerite (ZnS) demonstrates a resinous luster.

Pearly

Pearly luster appears iridescent, opalescent, or pearly. This is typically exhibited by mineral surfaces which are parallel to planes of perfect cleavage. Layer silicates such as talc often demonstrate a pearly luster on cleavage surfaces.

Greasy

A surface which possesses greasy luster appears to be covered with a thin layer of oil. A light-scattering surface which is slightly rough, such as that of nepheline, may exhibit greasy luster.

Silky

Silky luster occurs when light is reflected off of an aggregate of fine parallel fibers; malachite and serpentine may both exhibit silky luster.

Vitreous

Vitreous luster occurs in minerals with predominant ionic bonding and resembles the reflective quality of broken glass. The refractive index of such minerals is 1.5 to 2.0. Many silicates possess this type of luster; quartz and tourmeline both demonstrate vitreous luster.

Adamantine or Brilliant

A brilliant luster such as the sparkling reflection of diamond is known as adamantine. Minerals of adamantine luster have high refractive indices (1.9-2.6) and are highly dispersive and translucent. Covalent bonding or the presence of heavy metal atoms or transition elements may result in adamantine luster.

Density

The property of density is defined as mass per unit volume:

$$\mu = m/V$$

The geometric structure of the unit cell of a mineral determines the volume which it occupies. The masses of the atoms which compose the unit cell decree the mass of each cell. The identity of the atoms which compose the unit cell is specified by the chemical formula of the mineral. Density is therefore directly related to both the physical structure of the unit cell and the chemical composition of each species of mineral.

One method of measuring the density of a sample entails the use of one dense liquid and another miscible liquid of lower density. A solution of the two substances is created in which a crystal of the mineral in question remains suspended and neither sinks nor floats. The weight of a known volume of the solution is then measured, and the density of the solution and thus the density of the crystal are calculated from this information. Bromoform ($CHBr_3$, density 2.9 g/cm³), soluble in acetone; di-iodomethane (CH_2I_2, density 3.3 g/cm³), soluble in chloroform, $CHCl_3$; and Clerici's solution (a solution of thallium formate and thallium malonate; density 4.4 g/cm³), soluble in water, are some heavy liquids and their solvents which are commonly used in this process.

Density has historically been equated by mineralogists with the concept of specific gravity. Specific gravity is a unitless quantity which is defined as the ratio of the weight of a substance to the weight of an equal volume of water at a temperature of 4 °C. This ratio is equal to the ratio of the density of the substance to the density of water at 4 °C.

$$G = \mu / \mu_{water}$$

Specific gravity has therefore classically been measured by weighing a mineral specimen on a balance scale while it is submerged first in air and then in water. The difference between the two measurements is the weight of the volume of water which was displaced by the sample. The specific gravity of the mineral specimen is thus:

$$G = m_{air} / [m_{air} - m_{water}]$$

Because the density of water at 4° Celsius is 1.00 g/cm³, the density of a mineral in units of grams per centimeter cubed (g/cm³) is equal to its (unitless) specific gravity.

The field geologist sometimes uses a very rough estimation of the density of a hand-held sample as a clue to identification. Certain rough trends relating mineral density to various other factors are sometimes useful. Native elements, which contain only one type of atom and whose molecular structure is that of cubic or hexagonal closest packing, are relatively dense. Minerals whose chemical composition contains heavy metals - atoms of greater atomic number then iron (Fe, atomic number 26) - are more dense than atoms whose chemical composition does not include such elements. Minerals which formed at the high pressures deep within the earth's crust are in general more dense than minerals which formed at lower pressures and shallower depths. A general trend relating color to density is also prevalent; this trend states that dark-colored minerals are often fairly heavy whereas light-colored ones are frequently relatively light. A geologist is thus given cause to remark upon a sample which seems to reverse this trend. For example, graphite is dark colored but of low density (C; 2.23 g/cm³) while barite is light in color but unexpectedly heavy (BaSo$_4$; 4.5 g/cm³). The noted oddity of unexpectedly high or low density with respect to color provides the field geologist with a clue as to the identification of such atypical materials.

Hardness

Hardness has traditionally been defined as the level of difficulty with which a smooth surface of a mineral specimen may be scratched. The hardness of a mineral species is dependent upon the strength of the bonds which compose its crystal structure. Hardness is a property characteristic to each mineral species and can be very useful in identification.

Certain trends exist in hardness with respect to mineral class. Native elements are typically soft, although iron (Fe) and platinum (Pt) are relatively hard and diamond (C) is exceptionally hard. Compounds of heavy metals are soft. Sulphides and sulpho-salts, with the exception of pyrite, are relatively soft; halides are soft; carbonates and sulphates are usually soft. Oxides are typically hard while hydroxides are softer. Anhydrous silicates tend to be hard, while hydrous silicates are softer.

The Mohs Scale

The property of hardness has historically been measured according to the Mohs scale, which was created in 1824 by the Austrian mineralogist Friedrich Mohs. Mohs based his system for measuring and describing the hardness of a sample upon the definition of hardness as resistance to scratching. Mohs' method thus relies upon a scratch test in order to relate the hardness of a mineral specimen to a number from the Mohs scale.

In order to define his scale, Mohs assembled a set of common reference minerals of varying hardnesses and labled these in order of increasing hardness from 1 to 10. The reference minerals of the Mohs scale are as follows:

1. Talc
2. Gypsum
3. Calcite
4. Fluorite
5. Apatite

6. Orthoclase

7. Quartz

8. Topaz

9. Corundum

10. Diamond

Each reference mineral will scratch a test specimen with a Mohs hardness less than or equal to its own. Each reference mineral can be scratched by a specimen with a hardness equal to or greater than its own. If a reference mineral both scratches and can be scratched by a certain test specimen, then the specimen is assumed to possess a hardness equal to that of the reference mineral in question.

The set of reference minerals of the Mohs' scale can be supplemented by a few common household items. A fingernail has a Mohs hardness of 2½; a copper penny 3, window glass 5½, and a knife blade approximately 6.

The hardness of an unknown sample can be determined to within ½ increment by using the scratch test. Mineral hardnesses determined by the scratch test should never be given in decimal form, because the Mohs scale does not provide measurements of such precision.

The hardness of a mineral may vary with direction and crystallographic plane. This effect is usually small. However, species exist in which the variance in the hardness along different axes is notable. For example, the mineral kyanite (Al_2OSiO_4) typically forms elongated crystals. The Mohs hardness parallel to the length of a kyanite crystal is 5, whereas the Mohs hardness perpendicular to the length of such a crystal is 7. A second example is provided by the mineral halite, which is softer parallel to its cleavage planes than it is at a 45° angle to the cleavage planes.

Diamond Indentation Method

Investigations more recent than those completed by Mohs have used the diamond indentation method to quantitatively determine hardness. According to this method, a diamond point is pushed into a planar mineral surface under the weight of a known load. The diameter of the indentation thereby produced is then measured under a microscope. The diamond indentation hardness of a sample is equal to the mass of the load applied divided by the surface area of the indentation produced. The units in which diamond indentation hardness is recorded are therefore kilograms per millimeter squared (kg/mm^2).

Tests utilizing the diamond indentation method have shown that in order for a point fashioned from a certain material to scratch a surface the hardness of its constituent material must be 1.2 times that of the surface. Thus on an ideal hardness scale, each subsequent reference material would have a hardness of approximately 1.2 times that of the material preceeding it. It must be noted that the intervals between reference points on the Mohs scale are not, in fact, equal. The interval between subsequent reference points on the scale increases as the hardness of the reference materials increases. The skill with which Mohs chose his reference materials becomes apparent when one notes that each of his samples is approximately 1.6 times the hardness of the last.

The Mohs scale provides a means of testing hardness which is far more readily available to amateur geologists than the diamond indentation method. It has therefore remained the standard scale by which hardness is measured.

Cleavage

A cleavage plane is a plane of structural weakness along which a mineral is likely to split smoothly. Cleavage thus refers to the splitting of a crystal between two parallel atomic planes. Cleavage is the result of weaker bond strengths or greater lattice spacing across the plane in question than in other directions within the crystal. Greater lattice spacing tends to accompany weaker bond strength across a plane, because such bonds are unable to maintain a close interatomic spacing.

Both the positioning of crystal faces in a mineral and the property of cleavage are derived from the crystalline structure of the species. However, despite the fact that every mineral belongs to a specified crystal system, not every mineral exhibits cleavage. A mineral such as quartz may demonstrate beautiful, well-developed crystals and yet possess no distinct planes of cleavage.

Cleavage planes, if they exist, are always parallel to a potential crystal face. However, such planes are not necessarily parallel to the faces which the crystal actually displays. Fluorite, for example, has octahedral cleavage yet forms cubic crystals. Nonetheless, the property of cleavage, if it is present, can offer important information about the symmetry and inner structure of a crystal.

The quality of a mineral's cleavage refers to both the ease with which the mineral cleaves and to the character of the exposed cleavage surface. The quality of a sample's cleavage is typically described by terms such as 'eminent,' 'perfect,' 'distinct,' 'difficult,' 'imperfect,' or 'indistinct.'

'Eminent' cleavage describes the case in which cleavage always occurs readily and is in fact difficult to prevent from occurring. The mineral mica, for example, cleaves readily into thin, flat sheets. A mineral which demonstrates 'perfect' cleavage breaks easily, exposing continuous, flat surfaces which reflect light. Fluorite, calcite, and barite are minerals whose cleavage is perfect. 'Distinct' cleavage implies that cleavage surfaces are present although they may be marred by fractures or imperfections. 'Difficult' or 'indistinct' cleavage produces surfaces which are neither smooth nor regular; samples possessing such cleavage tend to fracture rather than split.

Cleavage may be determined by the examination of surfaces which have actually broken. It may also be determined by inspection of the interlacing systems of cracks which permeate the structure of certain specimens. These systems of cracks are beautifully apparent within transparent crystals such as fluorite or calcite.

Fracture

A mineral fractures when it is broken or crushed. Fracture takes place when a mineral sample is split in a direction which does not serve as a plane of perfect or distinct cleavage. In other words, fracture takes place along a plane possessing difficult, indistinct, or nonexistent cleavage. The difference between fracture and indistinct cleavage is not clearly delineated.

Unlike perfect or distinct cleavage, fracture does not result in the emergence of clearly demarcated planar surfaces which run parallel to possible crystal faces. Fracture is nondirectional: minerals which do not possess distinct cleavage may fracture in any possible direction.

Fractured surfaces may in some minerals possess a characteristic appearance which can aid in identification. Examples of distinctive types of fracture are 'conchoidal,' 'irregular,' and 'hackly' fracture.

Conchoidal

Conchoidal fracture results in a series of smoothly curved concentric rings about the stressed point, generating a shell-like appearance. The familiar ripples of a broken glass bottle demonstrate this type of fracture. Quartz and olivine are two mineral species which possess conchoidal fracture.

Irregular

Irregular or uneven fracture results in a rough, rugged surface.

Hackly

The term 'hackly' describes a fractured surface with multiple small, sharp and jagged irregularities.

Tenacity

The property of tenacity describes the behavior of a mineral under deformation. It describes the physical reaction of a mineral to externally applied stresses such as crushing, cutting, bending, and striking forces. Adjectives used to characterize various types of mineral tenacity include 'brittle,' 'flexible,' 'elastic,' 'malleable,' 'ductile,' and 'sectile'.

Brittle

Most mineral species are brittle, and will crumble or fracture under pressure or upon the application of a blow. Such materials break or powder easily.

Flexible

A mineral which is flexible rather than brittle will flex as opposed to breaking under the application of stress. However, a mineral which is merely flexible and not also elastic will be unable to return to its original shape when the stress is removed. Flakes of molybdenite and scales of talc are two substances which are flexible but inelastic.

Elastic

An elastic mineral will deform under external stress but will resume its original shape after the stress is removed. If it is bent, it will flex, but will return to its previous position when the stress disappears. The mineral called mica is both flexible and elastic.

Malleable

Native metals such as copper, silver, and gold are easily flattened with a hammer. This type of tenacity is termed malleable. Metallic-bonded minerals tend to be malleable, and may be pounded out into thin, flat sheets.

Ductile

Some malleable materials are also ductile, and may be drawn out into a thin wire without crumbling.

Sectile

Some minerals may be sliced into smooth sheets with a knife, although these may possibly still crumble under a blow from a hammer. Materials possessing this rare type of tenacity are called sectile minerals. The species chlorargyrite (AgCl) offers an example of a sectile mineral.

Crystal Habit

The term crystal habit describes the favored growth pattern of the crystals of a mineral species, whether individually or in aggregate. It may bear little relation to the form of a single, perfect crystal of the same mineral, which would be classified according to crystal system. Subtle evidence of the crystal system to which a mineral species belongs is, however, frequently observed in the habit of the crystals which a specimen displays.

The terminology used to describe crystal habit is not intended to replace the precise nomenclature of crystallography. Instead, it is intended as a supplement to this system. Discussions of crystal habit are more descriptive than precise; for this reason the terminology is suited to the discussion of mineral samples discovered in the field. Naturally formed specimens are rarely quantitatively perfect.

The crystals of particular minerals species sometimes form very distinctive, characteristic shapes. Crystal habit is thus often useful in identification.

Although each mineral species typically forms according to a few preferred shapes, crystal habit is largely determined by the environmental conditions under which a crystal develops. For example, aqueous solutions near or surrounding a crystal contain the elemental substances which it needs to continue growth. The direction from which a growing crystal may obtain such solutions is a factor which will affect its eventual shape. Higher environmental temperatures during formation increase ion mobility and aid in crystal formation; the rate at which the environment cools determines how much time a mineral is allowed to form large crystals. The amount of space available for a crystal to fill affects its final shape and size. Surface energy relations are also quite important to the direction of crystal growth; this process is not yet fully understood.

Electrical Properties

Three electrical properties are applicable to minerals: Conduction, Pyroelectricity, and Piezoelectricity.

Conduction

Conduction in mineral terms is defined as the ability of a mineral to conduct electricity. Only a very small number of minerals are good conductors; they are the metallic elements and the mineral

Graphite. These conductors can be placed between a wire carrying electricity, and the electricity will pass through. Conduction is an important property that can distinguish true metals from metallic looking sulfides and oxides.

Pyroelectricity

Pyroelectricity describes the ability of a mineral to develop electrical charges when exposed to temperature changes. Some minerals develop an electrical charge when heated, others when cooled.

Piezoelectricity

Piezoelectricity describes the ability of a mineral to develop electrical charges when put under stress. Piezoelectric minerals will develop charges when rubbed or struck repeatedly.

Conduction is very useful in distinguishing true metals, but pyroelectricity and piezoelectricity are not practical testing methods for normal mineral collectors.

Optical Properties

Several important optical properties are applicable to minerals and gemstones, and can be very useful for gem identification. With proper equipment, jewelers can easily distinguish a Ruby from Garnet or red glass, even if their outward appearance may be identical.

White Light, or visible light, is a form of electromagnetic radiation (energy waves produced by the motion of an electric charge). White light belongs to the color spectrum, which defines all forms of light and electromagnetic radiation. The spectrum also includes many forms of light not visible to the human eye, such as ultraviolet and infrared light.

The rate of motion produced by the electrical charge defines the wavelength of light. Different wavelengths produce different types of light; white light encompasses all wavelengths visible to the human eye. White light contains the seven primary colors: red, orange, yellow, green, blue, indigo, and violet.

The color white is a composite of all colors. Every substance receives its color from the way white light reacts to it. Light can either be absorbed into a substance, or it can be reflected. The presence of certain elements or chemicals in a substance determine which wavelengths (i.e. colors) are reflected and which are absorbed. The wavelengths reflected off a substance determine its color. For example, if a substance absorbs all wavelengths except for yellow and green wavelengths, which are reflected, the color of the substance is yellow-green. If a substance absorbs no wavelengths but reflects them all, its color is white. If, however, a substance absorbs all wavelengths, its color is black.

In addition to reflection and absorption, light can also be passed through a substance. Light passing through a substance determines its transparency. If all light passes through a substance, and none is reflected or absorbed, the substance is transparent and colorless. These three attributes (reflection, absorption, and passing through) determine the color and transparency of a substance.

Refractive Index (RI)

The speed of light varies in substances. The speed of light is different in air, water, and other dimensions, including minerals and gemstones. When light travels from one dimension to another dimension, the light bends, or refracts, upon entering the second dimension. This phenomenon can be witnessed with a stick protruding from a pond, where the stick appears to "bend" at the water level. This is caused by the difference in the speed of light in air and the speed of light in water. How much the light will bend, or the angle of refraction, depends on difference in the speed of light of the two substances. All transparent gemstones refract light, since the speed of light is different in the air than in gemstones.

Every gemstone refracts at a distinct, individual angle. The angle of refraction is directly related the speed of light in the gemstone. The refractive index of a gemstone measures the difference between the speed of light in air and the speed of light in the gemstone. This is determined by the gemstone's angle of refraction. Every gemstone has a unique refractive index, meaning every gemstone refracts light at a unique angle.

The refractive index value measures how much slower light travels in the gemstone than in the air. For example, the refractive index of Diamond is 2.42. This means that the speed of light in Diamond is 2.42 times slower than the speed of light in air.

Refractive indices of minerals range from 1.2 to about 3. However, gemstones with a refractive index greater than Diamond (2.42) are either synthetic or are too soft for practical gemstone use. The greater the refractive index of a gemstone, the more brilliant or lustrous it is.

The refractive index of a gemstone is measured with a refractometer, a tool that measures angle of refraction. This tool is used by almost all gemologists and professional jewelers, for it provides simple, inexpensive, and accurate gem identification. However, a refractometer cannot read values greater than 1.86. Gemstones with refractive indices greater than 1.86 can only be tested by placing them in a liquid with a known refractive index, and then calculating the difference in refraction between the liquid and the gem.

Double Refraction (DR)

Another optical property, known as double refraction or birefringence, is present in all non-amorphous minerals that do not crystallize in the isometric crystal system. When light rays enter birefringent minerals (minerals with double refraction), the light divides into two rays. The two rays differ in their angle of refraction. Therefore, all birefringent minerals have two refractive indices, one for each ray. The double refraction in most minerals is so weak that it cannot be observed with the naked eye. However, a small number of minerals have a strong double refraction, which is easily seen when the crystal placed over an image appears to "double" the image.

Double refraction is an important guide to gem identification. When viewed through a refractometer, birefringent minerals show two readings – one for each refracted ray of light. Double refraction is a characteristic trait, meaning every specimen of the same gem always has the same double refraction.

Double refraction is measured by the difference of refraction in each light ray. For example, if a gem is placed in a refractometer and shows a double reading of 1.62 and 1.63, its double refraction is .01.

Highly birefringent gemstones, such as Zircon, must be cut in a way where the double refraction is least noticeable in the finished gem, for it otherwise appears blurry.

Dispersion and Fire

Dispersion is the splitting of white light into the colors of the spectrum. This effect is observable in faceted, transparent, colorless gems, where the white light disperses in the gem and reflects on its inner surfaces, giving the gem a colorful sparkle. This effect is known as fire in the gem trade. Generally, gems with higher refractive indices display greater fire. Diamond has the greatest fire of all true, non-synthetic gems. Room lighting conditions play an important role in fire, for the stronger the light, the more intense the fire appears to be.

The design of the brilliant cut was extensively researched to offer a gemstone its maximum amount of fire. For this reason, many transparent gems with high dispersion are faceted with the brilliant cut.

Absorption Spectrum

The absorption spectrum describes the spectral wavelengths absorbed by a gem. The chemical structure of each gem allows only certain wavelengths to be absorbed (the rest are reflected or pass through). The wavelengths absorbed into a gem can be detected with an instrument known as a spectroscope. Color alone cannot be used to identify a gem, for many gems have identical colors. A spectroscope examines the "true" color of a gem. Examining the absorption spectrum of a gem is one of the most useful and practical methods of gem identification.

Miscellaneous Properties

1. Magnetite: The mineral magnetite is magnetic. Lodestone, a variety of magnetite, is a natural magnet. To test for magnetism, suspend the magnet from a string and then draw it near the sample. If magnetic, the magnet will be deflected toward the sample.

2. Plagioclase feldspar: The mineral plagioclase feldspar commonly shows very small, parallel lines on some of the cleavage planes. These are called striations. In some cases, these striations are the only way to distinguish plagioclase feldspar from orthoclase feldspar.

3. Graphite: It has a greasy feel. After feeling it, check your fingers. They will be dirty since graphite also has a hardness of "1." You can also see if you can mark on paper with the sample. It is "pencil lead."

4. Sulfur: It is a native mineral with a chemical formula of S. It smells like rotten eggs when streaked or scratched. It is also bright yellow in color.

5. Calcite: Calcite will effervesce (fizz) when a drop of concentrated lemon juice is placed on its surface. It also has a hardness of 3- it can be scratched with a penny but not with your fingernail.

Magnetic Properties

Several minerals react when placed within a magnetic field. Some minerals are strongly attracted to the magnet, others are weakly attracted, and one mineral is actually repelled. There are also several minerals that are attracted to magnetic fields only when heated.

A magnetic field is an area encompassing a magnet or electrical current that has the ability to attract or repel certain objects placed in the field. The closer the object is to the magnet or electrical current, the more powerful the magnetic effect. In virtually all cases, the presence of the element iron as a component of the mineral's chemical structure is responsible its magnetic properties.

Magnetic properties of minerals are defined as follows:

1. Ferromagnetism describes strong attraction to magnetic fields. This property is exhibited in few minerals, notably Magnetite and Pyrrhotite.

2. Paramagnetism is weak attraction to magnetic fields. The attraction is usually discernible, but it may be so weak that it is undetectable. Most paramagnetic minerals become strongly magnetic when heated. A small number of paramagnetic minerals, such as Platinum, are not essentially paramagnetic, but contain iron impurities which are responsible for the paramagnetism. However, some specimens lacking iron also exist, and these are not paramagnetic. Some examples of paramagnetic minerals are Hematite and Franklinite.

3. Diamagnetism. Only one mineral, Bismuth, is diamagnetic, meaning it is repelled from magnetic fields.

4. Another property, which is unnamed, is attraction to magnetic fields when heated. Some iron sulfides and oxides become ferromagnetic after heating, as a result of combined sulfur or oxygen ions freeing themselves from the iron. Some minerals may even act as magnets when heated.

Only a variety of one mineral acts as a magnet, generating magnetic fields on its own. This mineral is Lodestone, the magnetic variety of Magnetite, which found in only a few deposits throughout the world. Although it is only weakly magnetic, its magnetism is definitely discernible.

Lodestone: The magnetism of this magnetic variety of Magnetite is clearly visible.

Magnetic properties are useful for identifying a mineral, for if observed it can pinpoint a mineral. The most effective testing results are obtained with the use of a powerful magnet. The only minerals that possibly respond to magnets without heating are opaque, metallic-looking minerals.

Silicate Minerals

The mineral quartz (SiO_2) is found in all rock types and in all parts of the world. It occurs as sand grains in sedimentary rocks, as crystals in both igneous and metamorphic rocks, and in veins that cut through all rock types, sometimes bearing gold or other precious metals. It is so common on Earth's surface that until the late 1700s it was referred to simply as "rock crystal." Today, quartz is what most people picture when they think of the word "crystal."

Quartz falls into a group of minerals called the silicates, all of which contain the elements silicon and oxygen in some proportion. Silicates are by far the most common minerals in Earth's crust and mantle, making up 95% of the crust and 97% of the mantle by most estimates. Silicates have a wide variety of physical properties, despite the fact that they often have very similar chemical formulas. At first glance, for example, the formulas for quartz (SiO_2) and olivine (($Fe,Mg)_2SiO_4$) appear fairly similar; these seemingly minor differences, however, reflect very different underlying crystal structures and, therefore, very different physical properties. Among other differences, quartz melts at about 600 °C while olivine remains solid to temperatures of nearly twice that; quartz is generally clear and colorless, whereas olivine received its name from its olive green color.

The variety and abundance of the silicate minerals is a result of the nature of the silicon atom, and even more specifically, the versatility and stability of silicon when it bonds with oxygen. In fact, pure silicon was not isolated until 1822, when the Swedish chemist Jöns Jakob Berzelius finally succeeded in separating silicon from its most common compound, the silicate anion $(SiO_4)^{4-}$. This anion takes the shape of a tetrahedron, with an Si^{4+} ion at the center and four O^{2-} ions at the corners thus, the molecular anion has a net charge of -4.

Three ways of drawing the silica tetrahedron: a) At left, a ball & stick model, showing
the silicon cation in orange surrounded by 4 oxygen anions in blue; b) At center,
a space filling model; c) At right, a geometric shorthand.

The Si-O bonds within this tetrahedral structure are partially ionic and partially covalent, and
they are very strong. Silica tetrahedra bond with each other and with a variety of cations in many
different ways to form the silicate minerals. Despite the fact that there are many hundreds of
silicate minerals, only about 25 are truly common. Therefore, by understanding how these silica
tetrahedra form minerals, you will be able to name and identify 95% of the rocks you encounter on
Earth's surface.

Structure of the Silicates

Early mineralogists grouped minerals according to physical properties, which spread the silicates
across many groups because they have very different properties. By the early 1800s, however,
Berzelius had begun classifying minerals based on their chemical composition rather than on their
physical properties, defining groups such as the oxides and sulfides – and, of course, the silicates.
At the time, Berzelius was able to determine the absolute proportions of elements within a mineral,
but he could not see the internal arrangement of the atoms of those elements in their crystalline
structure.

A detailed view of the internal arrangement of atoms within minerals would have to wait over
100 years for the development of X-ray diffraction (XRD) by Max von Laue, and its application to
determine atomic distances by the father-son team of William Henry Bragg and William Lawrence
Bragg a few years later. In the process of XRD, X-rays are aimed at a crystal. Electrons in the atoms
within the crystal interact with the X-rays and cause them to undergo diffraction. In the same way
that light can be diffracted by a grate or card, X-rays are diffracted by the crystal and a 2-dimen-
sional pattern of constructive and destructive interference bands results. This pattern can be used
to determine the distance between atoms within the crystal structure according to Bragg's Law.
The Braggs' work opened up a new world of mineralogy, and they were awarded a Nobel Prize in
1915 for their work determining the crystal structures of NaCl, ZnS, and diamond. XRD revealed
that even minerals with similar chemical formulas could have very different crystal structures,
strongly influencing those minerals' chemical and physical properties.

As scientists created XRD images of the atomic structure of minerals, they were better able to understand the nature of the bonds between atoms in the silicate and other crystals. Within a silica
tetrahedron, any single Si-O bond requires half of the available bonding electrons of the O^{2-} ion,
meaning that each O^{2-} may bond with a second ion, including another Si^{4+} ion. The result of this
is that the silica tetrahedra can polymerize, or form chain-like compounds, by sharing an oxygen
atom with a neighboring silica tetrahedron. The silicates are, in fact, subdivided based on the

shape and bonding pattern of these polymers, because the shape influences the external crystal form, the hardness and cleavage of the mineral, the melting temperature, and the resistance to weathering. These different atomic structures produce recognizable and consistent physical properties, so it is useful to understand the structures at an atomic level in order to identify and classify the silicate minerals. Identifying minerals in a rock may seem like an arcane exercise, but it is only by identifying minerals that we begin to understand the history of a given rock.

The most common silicate minerals fall into four types of structures, described in more detail below: isolated tetrahedra, chains of silica tetrahedra, sheets of tetrahedra, and a framework of interconnected tetrahedra.

Isolated Tetrahedra: Olivine

The simplest atomic structure involves individual silica anions and metal cations, usually iron (Fe) and magnesium (Mg), both of which exist most commonly as ions with charge of +2. Therefore, it takes two atoms of Fe^{2+} or Mg^{2+} (or one of each) to balance the -4 charge of the silica anion. Olivine is the most common silicate of this type, and it makes up most of the mantle. Because these minerals contain a relatively high proportion of iron and magnesium, they tend to be both denseand dark-colored. Because the tetrahedra are not polymerized, there are no consistent planes of internal atomic weakness, so they also have no cleavage. Garnet is another common mineral with this structure.

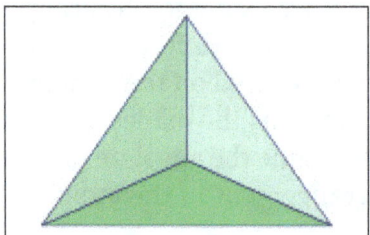

Depiction of a single silicate tetrahedron.

A picture of olivine (the green crystals), an example of a silicate
structure composed of isolated tetrahedrons, with a vein of basalt (the gray material).

Chains of Tetrahedra: Pyroxenes and Amphiboles

When silicate anions polymerize, they share an oxygen atom with a neighboring tetrahedron. Commonly, each tetrahedron will share two of its oxygen atoms, forming long chain structures. These chains still have a net negative charge, however, and the chains bond to metal cations like Fe^{2+}, Mg^{2+}, and Ca^{2+} to balance the negative charge. These metal cations commonly bond to multiple chains, forming bridges between the chains. Single-chain silicates include a common group called the pyroxenes, which are generally dark-colored. Because the bonds within the tetrahedra are strong, planes of atomic weakness do not cross the chains; instead, pyroxenes have two cleavage planes parallel to the chains and at nearly right angles to each other.

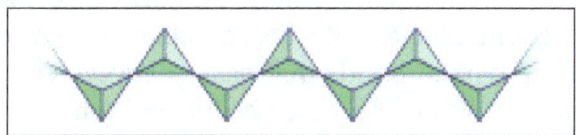

A schematic diagram of the single chain silica structure. Where two tetrahedra touch, they share an oxygen ion.

Pyroxene is one of the dominant minerals in this sample of gabbro. It is the dark mineral and can be hard to recognize.

Double chains form when every other tetrahedron in a single chain shares a third oxygen ion with an adjoining chain. Like single chains, the double chains still maintain a net negative charge and bond to cations that can form bridges between multiple double chains.

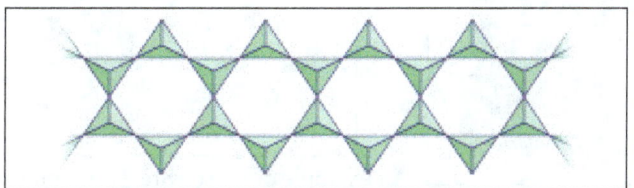

A schematic diagram of the double chain silicate structure.

Double chain silicates, called amphiboles, host a wider variety of cations, including Fe^{2+}, Mg^{2+}, Ca^{2+}, Al^{3+}, and Na^+, and have a wide variety of colors. The most common amphibole is hornblende, a black mineralfound in igneous rocks like granite and andesite. Amphiboles tend to form prismatic crystals with two cleavage planes at 120 degrees to each other.

Individual hornblende crystals where the characteristic cleavage can be seen.

Hornblende is the dark mineral in this rock.

Pyroxenes and amphiboles can be difficult to distinguish from one another, as they are both dark-colored, blocky minerals. A careful examination of the angle between cleavage planes, described above, is required to identify them.

Sheets: Micas and Clays

When every tetrahedron shares three of its oxygen ions with neighboring tetrahedra, sheets are formed. Micas such as muscovite and biotite are both common sheet silicates, notable for their one perfect cleavage. This perfect cleavage results from the type of bonds that occur between sheets – van der Waals bonds. Because van der Waals bonds are weak, cleavage occurs between sheets, never across sheets. Clays are another very important sheet silicate that incorporate water into their atomic structure. The presence of water lubricates the sheets and is what makes clays easy to work with in forming pottery; the firing process heats the minerals to the point where the water is driven off, resulting in a rigid, durable structure such as a pot.

An example of biotite.

An example of muscovite. (Both biotite and muscovite are micas, which are one kind of sheet silicate.)

Framework: Quartz and Feldspar

When each tetrahedron shares all of its oxygen atoms with adjacent tetrahedra, a very strong 3-dimensional framework of Si-O bonds is formed. Quartz is pure SiO^2; note that the charge is now exactly balanced and no other bonding ions are needed. In the feldspars, one or two out of every four Si^{4+} ions is replaced by an Al^{3+} ion, creating a charge imbalance that must be solved through the presence of additional cations: K^+, Na^+, and Ca^{2+}. There are two kinds of feldspars upon which cations are incorporated into the structure. Feldspars that contain the K^+ cation are called K-feldspars, or alkali feldspar, whereas those that contain Na^+ and Ca^{2+} are called plagioclase feldspars. This separation occurs because K^+ is a much larger cation than either Ca^{2+} or Na^+, and its presence creates a slightly expanded framework structure.

An example of the 3-dimensional structure formed by a framework silicate.

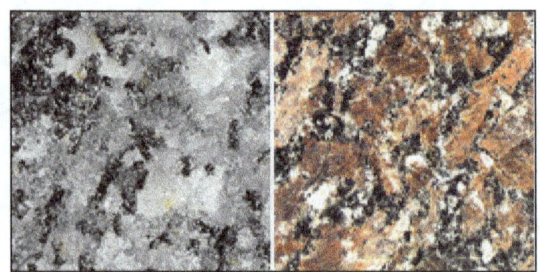

The white, blocky minerals in the rock on the left are plagioclase feldspar; the pink minerals in the rock on the right (granite) are K-feldspar.

Like olivine, quartz also has no cleavage, because there is no natural weakness within that 3-dimensional framework. The feldspars, on the other hand, have two good cleavage planes at ~90 degrees to each other, due in part to the way that the aluminum ion changes the structure slightly, opening up planes of weakness. Quartz and feldspar are generally light-colored as well, making them easily distinguishable from darker minerals like olivine and pyroxene.

Quartz and feldspar together make up the bulk of the rocks we see at the surface. Plagioclase feldspar is the single most common mineral in Earth's crust, making up an estimated 39% of both continental and oceanic crust. Quartz only makes up an estimated 12% of the entire crust, but it is by far the most common mineral we see on the surface because of its resistance to weathering.

Familiarity with these few minerals – olivine, garnet, pyroxene, hornblende, muscovite, biotite, K-feldspar, plagioclase, and quartz – prepares you to identify and interpret the vast majority of rocks you will see on Earth's surface.

Silicates as a Natural Resource

Though we generally think of coal or oil when discussing natural resources, silicate minerals are a natural resource we can't live without on our planet, and not just because of our increasing reliance on computers. Without quartz, there would be no glass. Without the clay minerals, we would have no ceramics or pottery. We use silicate minerals in the manufacture of many building materials, including bricks and concrete. The weathering of silicate minerals on the surface of Earth produces the soils in which we grow our foods and the sand on our beaches. The properties of the minerals that are important to us are based on the versatility of the silicate anion in combination with other elements.

Oxide Minerals

Oxide minerals can be listed as compounds of oxygen with metals. A list of the common oxide minerals with the spinel structure, together with their compositions, u values, cell dimension and structure type. Pure end member compositions of these minerals arc rare in nature, and cation substitution, especially between cations of similar size and valence, results in extensive solid solutions between them. An important solid solution is that between magnetite Fe_3O_4 and Fe_2TiO_4 as it is the main carrier of magnetism in rocks. A solid solution implies random substitution of cations for one another i.e. disorder. We have already seen in general terms that disorder at high temperatures tends to give way to order at lower temperatures.

Magnetite or Magnetic Iron Ore: Fe_3O_4

Mineral Name	Composition
Magnetite	$Fe^{3+}[Fe^{2+}Fe^{3+}]O_4$
Magnesioferrite	$Fe^{3+}[Mg^{2+}Fe^{3+}]O_4$

Chromite	$Fe^{2+}[Cr_2^{3+}]O_4$
Magnesiochromite	$Mg^{2+}[Cr_2^{3+}]O_4$
Spinel	$Mg^{2+}[Al_2^{3+}]O_4$
Hercynite	$Fe^{2+}[Al_2^{3+}]O_4$
Ulvospinel	$Fe^{2+}[Fe^{2+}, Ti^{4+}]O_4$
Jacobsite	$Fe^{3+}[Mn^{2+}, Fe^{3+}]O_4$

Magnetite and magnetic iron ore has for color, iron-black; powder, black; luster, metallic, but sometimes rather dull; brittle; H = 5.5 to 6.5; G = 5.168 to 5.180; composition (as calculated from the composition formula) iron, 72.4%, oxygen, 27.6%. Sometimes contains titanium, sometimes manganese. Crystals in the cubic system, mostly octahedrons, but often breaks with smooth, at surfaces (parting planes); strongly magnetic. The word magnetic is used in two senses: (a) attracted by a magnet; (b) attracts certain substances (as iron), i.e., it is itself a magnet. Magnetite is always magnetic in the rst sense; it is occasionally magnetic in the second sense, in which case, it is called loadstone. (Try its attraction on small pieces of iron.) A piece of loadstone has a north pole and a south pole, like a compass, which property is called polarity. The magnetic property of minerals can be conveniently tested with a magnetized knife blade; and a blade that has been rubbed with a strong magnet will keep its charge for many years. To test for magnetism, break off a number of small grains of the mineral, and bring the point of the magnetized knife blade to them slowly; if strongly magnetic, the grains will jump to the knife blade. Do not put the point of the blade on the little pieces, as they may stick together by reason of other causes besides magnetism. When testing for hardness by scratching the mineral with the point of the blade, a magnetic mineral will form a fringe around the point. But caution is necessary here; if the mineral is about the same hardness as the blade or harder, the fringe may be made up of bits of steel from the blade.

Magnetite is a common ore. In some localities, as in Sweden, it is valued on account of its purity, being free from phosphorus and sulphur; but, because it is usually harder and less porous than hematite, the latter is preferred, as it is easier to melt. Value, at a port near the furnace; but an ore low in phosphorus and containing 68% of iron is quoted higher.

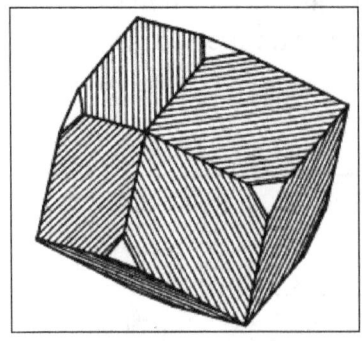

Magnetite crystal

Taking the specic gravity of the ore as 5, the number of cubic feet per ton is calculated as follows: weight of a cubic foot of water is 62.4 pounds; weight of a cubic foot of magnetite is 62.4 x 5 = 312 pounds; number of cubic feet to the ton is 2000 ÷ 312 = 6.41. Since a cubic yard contains 27 cubic feet, a cubic yard weighs 27 ÷ 6.41 = 4.21 tons. Henee, roughly, 6½ cubic feet weigh a ton, and a cubic yard weighs 4¼ tons. The ordinary ore, mixed more or less with rock, will run from 6½ to 7½ cubic feet to the ton.

Magnetic iron ore is found in igneous and metamorphic rocks in lenses or in vein-like deposits; also in beds of siliceous ore in pre-Cambrian rocks of sedimentary origin.

Ilmenite or Titanic Iron Ore: $FeTiO_3$

Ilmenite or Titanic Iron Ore are in Color, iron-black, very much like magnetite; powder, black to brownish red; luster, near-metallic; H = 5 to 6; G = 4.5 to 5. Composition: iron, 36.8%; titanium, 31.6%; oxygen, 31.6%. Very slightly magnetic, and thus distinguished from magnetite. Titanium in iron ores is objectionable, causing difculties in smelting; but since there are large deposits of magnetite high in titanium, though otherwise pure, these difficulties will, in time, be overcome, and these ores will then become valuable. Ilmenite is valued as an ore of titanium selling a ton of ore on the basis of 59% to 60% titanic oxide, TiO_2. Ilmenite is used for making titanium metal, as well as ferro-titanium, employed in making steel, and for making titanium white, a white pigment used for paints with great covering power and durability. Titanic oxide, made from ilmenite, is used to give a yellow color to enamel. In Norway, a number of paints (red, orange, yellow) have been manufactured by roasting titanic iron ores.

Found in gabbro, diorite, syenite, etc., often mixed with more or less magnetite. Sometimes altered to a dull-white substance, leucoxene, of variable composition, but sometimes of the same composition as titanite or rutile.

Franklinite

Franklinite is an oxide of iron, zinc, and manganese; has no simple composition formula, as the proportion of the metals vary. Color, iron-black; looks much like magnetites but is not so strongly magnetic; powder, reddish brown or black; H=5.5 to 6.5; G=5.07 to 5.22. An average analysis of franklinite shows it to contain iron, 45.53%, manganese, 10.29%, zinc, 18.70%, oxygen, 24.48%. It is used as an ore of zinc, and the residue as an iron ore; its value is not quoted on the market. Crystals, in the cubic system, octahedrons, like those of magnetite. Found with other zinc ores.

Hematite: Fe_2O_3

Hematite is of Color, dark steel-gray or iron black; powder, red or reddish brown; luster, metallic, sometimes dull; H = 5.5 to 6.5; G = 4.9 to 5.3. Hematite deposits are often compacted powder or partly so. The ore in such case is red. Red ocher is ne-grained hematite loosely compacted; reddle and red chalk are the same, with more or less clay; specular hematite is in brilliant black crystals (from speculum, a mirror); when the crystals are small and thin, like scales of mica, the mineral is micaceous hematite; clay ironstone is a hard, brownish-black to reddish-brown mixture of hematite and clay or sand, etc.; kidney ore is hematite in rounded masses, which break with a smooth fracture and more or less radiating structure.

All varieties of hematite give a red powder. Earthy and micaceous varieties seem much softer than the mineral really is, because the test results in separating the grains already formed; it does not give a correct idea of the hardness of the individual grains. Sometimes hematite is slightly magnetic. The pure mineral contains iron, 70%, oxygen, 30%. As mined, iron ores are more or less mixed with rock (gangue), which lessens the percentage of iron; an ore is considered rich if it carries 60% of iron, and the average per cent of iron in the ores, as fed into the furnaces is not much above 50. Poorer ores are often improved (beneciated) by washing, which removes the lighter material, or by magnetic concentration (in the case of low-grade magnetite).

Value of hematite is cheap for non-bessemer ore (that is, high in phosphorus), 51½% iron, and more for bessemer ore, 51½% of iron. Found in rocks of all ages, in vast quantities in sedimentary rocks, and in smaller masses in igneous rocks.

Limonite: $2F_2O_3 \cdot 3H_2O$

Limonite is in Color, brown or black, the compact, hard ore is dark brown, sometimes nearly black; when soft and earthy (bog ore and yellow ocher), it is brownish yellow or yellow; powder, yellow or yellowish brown; luster, non-metallic, near-metallic, or earthy, sometimes silky; H = 5 to 5.5, but the earthy varieties crumble easily when tested with the point of a knife, and thus may seem quite soft; G = 3.6 to 4; not crystalline. The composition is the same as that of hematite, with the addition of water (H_2O). The formula as written above shows that there are 2 (56 x 2 +16 x 3) = 2 x 160 = 320 parts of hematite and 3 (1 x 2 + 16) = 54 parts of water; and the proportion of the iron in the ore is 2 x 56 x 2/320 + 54 = 224/374 = 0.599 = 59.9%. Limonite is found in large deposits of hard ore, and also as a loose covering of the bottoms of shallow lakes and bogs.

Goethite: $Fe_2O_3 \cdot H_2O$

Goethite is like limonite, but is crystalline; it is a little heavier than limonite (G = 4 to 4.4), and is sometimes reddish in color; powder, brownish yellow to yellow. Turgite, $2F_2O_3 \cdot H_2O$ is like limonite in appearance, but gives a red powder, like hematite. In composition, it lies between the two, having some water, but less than goethite.

Limonite is peculiar among valuable minerals in the fact that in some places it is being deposited fast enough to renew the supply in a few years after it has been exhausted. It is said that at Radnor Forges, in Quebec, the bottoms of the shallow lakes were cropped every seven or eight years. As iron ores, limonite, goethite, and turgite have values similar to those of magnetite and hematite. Ore such as was mined originally at the Helen Mine, Michipicoten District Ontario, and more recently at Steep Rock Lake, Ontario, and on the vast new iron range in Labrador and New Quebec, ranks with the best hematite.

Chromite: $FeCr_2O_4$

Chromite of Color, iron black to brownish black; powder, brown; luster, metallic or near-metallic; sometimes slightly magnetic; looks like magnetite, but is usually duller in luster ; seldom strongly magnetic; H = 5.5; G = 4.32 to 4.57. Composition, oxide of iron and chromium; the latter metal gives the mineral its value for the manufacture of chromates and similar chemicals, and paints (chrome yellow, chrome orange, chrome red, chrome green). The metal chromium is used in making the chrome-nickel alloys mentioned below; also, in making stellite alloys, mentioned under smaltite. Value of ore long ton for ore containing 48% to 51% of chromium oxide, provided the iron content is low. High iron reduces the value. Chromite is found in serpentine rocks.

Stellite Alloys

Since Ontario produces an appreciable part of the world's cobalt ores and Quebec has deposits of chromite, Canada has the raw materials for making stellite alloys, manufactured at Deloro, in eastern Ontario, since 1915. While the original stellite contained only cobalt and chromium, a third metal, tungsten, has been added, the percentages being: cobalt, 50% to 60%; chromium, 40% to 30%; tungsten, 20% to 8%. The alloy is used for the manufacture of machine tools, cutlery, surgical and dental instruments, evaporating dishes, annealing dishes, ornamental work, valves, plumbing xtures, pens, and combs. The silvery white color, stainlessness, and great hardness (some varieties are harder than quartz) fit for these uses. By using stellite tools, the speeds of machine tools have been increased as much as 50%. More recently its heat resistance has made it useful in parts of the jet engine.

Nichrome

Nichrome is an alloy of nickel, chromium, and iron. It resists high temperatures and is used for the heating elements of toasters, irons, furnaces, etc.

Stainless Steel

Iron alloyed with chromium in sufcient proportion resists rusting and other corrosion better than ordinary steel. The resisting power is particularly marked when about 20% of chromium is used, and is still greater when nickel is added as well as chromium. A good deal of cutlery is now made of high-chromium steel. It is also used in parts of machinery that have to withstand corrosion. Plating with chromium is commonly used for automobile nishing and for bearings.

Tinstone or Cassiterite: SnO_2

Tinstone or Cassiterite are of color, brown or black, sometimes red, gray, yellow, or white; pure oxide of tin is white; powder, brownish, gray, or white; luster, adamantine (like diamond); H = 6 to 7; G= 6.8 to 7.1; crystals, tetragonal; when in crystals, the luster is sometimes very brilliant, but other samples are dull. Wood tin has a somewhat brous structure, and mixed brownish shades, like dry wood. Toad's eye tin is similar to wood tin.

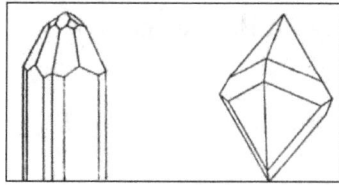
Tinstone Steel

Stream tin is pebbles and grains of tinstone, as found in beds of streams and on shores. Tinstone is the common ore of tin; its value is not often quoted; the value of the ore can be judged from the percentage of tin it contains. Tinstone is found in veins in granite, gneiss, mica schist, slaty schist, quartzite, and quartz porphyry. It is stated that in all productive tinstone areas, the ore is associated with a peculiar granite called greisen, which is composed mostly of quartz and white mica. The greater part of the world production of tin ore is stream tin.

Corundum: Al_2O_3

Corundum is of color, commonly gray or brown, also blue, red, yellow, and nearly white; pure aluminum oxide is white; powder, white or gray; luster, like diamond or glassy; H = 9; G = 3.95 to 4.10; fine, clear blue crystals are valuable as gems, and are then called sapphires; rubies are crystals of red corundum. Crystals, in hexagonal system; parting, good, but no true cleavage. Emery is usually black, fine- grained corundum, with magnetite or hematite; is sometimes hercynite $FeALO_4$.

Corundum being harder than quartz is valuable as an abrasive (material for grinding and polishing). Articial corundum is now made by an electric furnace process from noncrystalline aluminum oxide, which can be made cheaply from bauxite, and being of the same composition as corundum, requires only to be crystallized. Corundum is found in nepheline-syenite and similar rocks; also in peridotite.

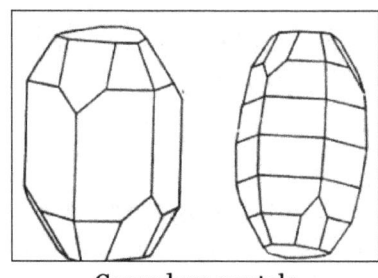
Corundum crystals

Bauxite: $Al_2O_3 \cdot 2H_2O$

Bauxite is of color, whitish, grayish, yellow, brown, red; powder, yellowish, brownish, or white; G = 2.55; earthy, often found with round masses like small eggs, and therefore called oolitic in structure. Composition is the same as corundum, with water in addition; but generally more or less limonite is mixed with the bauxite, causing the brown color. Its principal use is as an ore of aluminum. It is rst puried, then fed into a bath of fused cryolite, in which aluminum oxide is soluble; the aluminum is deposited by a current of electricity. Alundum is articial corundum made from bauxite by heating in electric furnaces. Bauxite bricks are used for furnace linings.

Bauxite is found in limestone and around the margins of clay deposits, in pockets or as "blanket" deposits.

Pyrolusite: MnO_2

Pyrolusite of the color, iron-black, dark steel-gray, sometimes bluish; powder, black or bluish black; luster, metallic; H = 2 to 2.5; soils the ngers when handled; G = 4.37 to 4.86; usually in long crystals, radiating or columnar, sometimes granular; composition, manganese oxide; it is often called manganese ore. Used in the manufacture of certain chemicals, particularly permanganates, but mostly in steel manufacture. Manganese steel, or Hadeld steel, usually contains 12% to 15% of manganese. It is very tough, and is used for jaws and other wearing parts of rock crushers and similar machinery; for railway frogs, crossings, and curves; for minecar wheels, trolley wheels, linings of ball mills, burglar-proof safes, etc. It lasts much longer than ordinary steel, because it is not only tough but becomes tougher with use. Manganese is added in small quantities to ordinary steel up to 1%; more than this makes the steel brittle; but when the amount of manganese is increased to 7%, the properties of Hadeld steel begin to appear. Ferro- manganese is an alloy of iron and manganese, containing from 40% to 80% manganese; it is used in making manganese steel. Spiegeleisen is pig iron very high in manganese, 15% to 30% ; it is used to give the required amount of manganese to ordinary steel.

Pyrolusite and other manganese ores are sold on the unit system, 90 to 95 cents per unit, the unit being 1% of manganese in a ton of ore. Pure pyrolusite contains about 65% of manganese. The best ore is used in chemical manufactures. Pyrolisite and other ores manganese are found in limestone, shale, sand- stone, etc., in veins and bunches; sometimes in quartzite, chert, and jasper; occasionally in bunches in clay.

Psilomelane or Hard Manganese Ore

Psilomelane or Hard Manganese Ore are in color, black, bluish, or brownish black; powder, brownish black, luster, silky or dull; H = 5 to 6; G = 4.1 to 4.7; composition, oxides of manganese with potash or baryta, with sometimes water; generally impure. This is a common ore of manganese; often found with pyrolusite, but distinguished from it by its greater hardness. Wad, or bog manganese is a brown, soft, earthy manganese ore, variable in composition ; it looks like bog iron ore, but is often darker, and is often found with pyrolusite. Asbolite, or earthy cobalt ore, resembles wad, but contains cobalt, sometimes enough to make it a cobalt ore. Lampadite is a variety of wad containing 3% to 13% of copper.

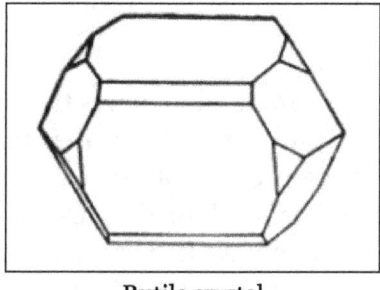
Rutile crystal

Braunite

Braunite are of color, brownish-black to steel-gray; powder the same; luster, near-metallic; H=6 to 6.5; G=4.75 to 4.82; crystals, tetragonal; cleavage, perfect; composition, oxides of manganese with some silica. This is an ore of manganese, often found with other manganese ores. Hausmannite, Mn_3O_4, is a similar mineral of brownish black color; powder, chestnut-brown; H =5 to 5.5; G=4.8.

Zincite or Red Zinc Ore; ZnO

The color of Zincite or Red Zinc is red to brownish red; powder, orange-red; luster, like diamond; H=4 to 4.5; G = 5.43 to 5.70; crystals, hexagonal, like quartz; cleavage, perfect; composition, zinc (about 80%) and oxygen. Pure oxide of zinc is white; the color of the mineral is caused by a little hematite or oxide of manganese. Zincite is an ore of zinc, but is chiey used in the manufacture of white zinc paint.

Cuprite or Red Copper Ore: Cu_2O

Colors of Cuprite or Red Copper Ore are of various shades of red sometimes almost black; powder, brownish red, shining; luster, somewhat metallic or earthly; H = 3.5 to 4; G = 5.85 to 6.15; crystals, in cubic system, usually octahedrons; cleavage, fair. Often has green covering of malachite; often looks like hematite, but is softer. Contains nearly 80% of copper. Found sometimes with the commoner ores of copper.

Rutile (M82): TiO_2

Rutile has a color reddish brown or red, sometimes yellowish, bluish, violet, black, or grassgreen; powder, pale brown; luster. somewhat metallic; H=6 to 6.5: G=4.18 to 4.25 (for a black variety, nigrine, G =5); crystals, in tetragonal system, like those of tinstone; cleavage, distinct; composition, titanium oxide, with a little iron, sometimes up to 10%. For uses, see Ilmenite. Value, is good for a pound for ore containing 94% titanium oxide (TiO_2). Found in granite, gneiss, mica slate, and syenite; sometimes in crystalline limestone and dolomite.

Carbonate Minerals

Carbonate minerals are any member of a family of minerals that contain the carbonate ion, CO_3^{2-}, as the basic structural and compositional unit. The carbonates are among the most widely distributed minerals in the Earth's crust.

The crystal structure of many carbonate minerals reflects the trigonal symmetry of the carbonate ion, which is composed of a carbon atom centrally located in an equilateral triangle of oxygen atoms. This anion group usually occurs in combination with calcium, sodium, uranium, iron, aluminum, manganese, barium, zinc, copper, lead, or the rare-earth elements. The carbonates tend to be soft, soluble in hydrochloric acid, and have a marked anisotropy in many physical properties (e.g., high birefringence) as a result of the planar structure of the carbonate ion.

There are approximately 80 known carbonate minerals, but most of them are rare. The commonest varieties, calcite, dolomite, and aragonite, are prominent constituents of certain rocks: calcite is the principal mineral of limestones and marbles; dolomite occurs as a replacement for calcite in limestones, and when this is extensive the rock is termed dolomite; and aragonite occurs in some recent sediments and in the shells of organisms that have calcareous skeletons. Other relatively common carbonate minerals serve as metal ores: siderite, for iron; rhodochrosite, for manganese; strontianite, for strontium; smithsonite, for zinc; witherite, for barium; and cerussite, for lead.

Most such rock-forming carbonates belong to one of two structure groups—either calcite or aragonite. The calcite structure is usually described with reference to the sodium chloridestructure in which the sodium and chloride of halite are replaced by calcium atoms and CO_3 groups, respectively. The unit cell of halite is distorted by compression along a three-fold axis, resulting in a rhombohedral cell. In calcite all CO_3 groups are parallel and lie in horizontal layers; CO_3 groups in adjacent layers, however, point in opposite directions. The calcium atoms are bonded to six oxygen atoms, one each from three CO_3 groups in a layer above and three from CO_3 groups in a layer below. The structure of dolomite, $CaMg(CO_3)_2$, is similar to that of calcite, $CaCO_3$, except that there is regular alternation of calcium and magnesium, and a lower symmetry, though still rhombohedral, results. The second structure group, that of aragonite, is orthorhombic. Like the calcite structure, the cation in the aragonite structure is surrounded by 6 carbonate groups; the CO_3 groups, however, are rotated about an axis perpendicular to their plane and the cation is coordinated to nine oxygen atoms instead of six.

Carbonate minerals other than simple carbonates include hydrated carbonates, bicarbonates, and compound carbonates containing other anions in addition to carbonate. The first two groups include nahcolite, trona, natron, and shortite; they typically occur in sedimentary evaporite deposits and as low-temperature hydrothermal alteration products. The members of the third group generally contain rare-earth elements and almost always result from hydrothermal alteration at low temperatures. Examples of these carbonate minerals are bastnäsite, doverite, malachite, and azurite.

References

- Mineral-chemical-compound, science: britannica.com, Retrieved 14 June, 2019

- Mineraloids, minerals: geology.com, Retrieved 15 July, 2019

- Physical-character, Mineralogy, bangert: mpp.mpg.de, Retrieved 16 August, 2019

- Electricalproperties, property, resource: minerals.net, Retrieved 17 January, 2019

- Optical, property, resource: minerals.net, Retrieved 18 February, 2019

- The-Silicate-Minerals, Earth-Science, library: visionlearning.com, Retrieved 19 March, 2019

- Carbonate-mineral, science: britannica.com, Retrieved 20 April, 2019

Mineralogy and its Branches

2

- **Mineralogy**
- **Optical Mineralogy**
- **Magnetic Mineralogy**
- **Environmental Mineralogy**
- **Clay Mineralogy**

The domain of geology which focuses on the scientific study of the crystal structure, chemistry and physical properties of minerals is known as mineralogy. A few of its branches are optical mineralogy, magnetic mineralogy, environmental mineralogy and clay mineralogy. This chapter has been carefully written to provide an easy understanding of these branches of mineralogy.

Mineralogy

Mineralogy is the branch of geology concerned with the study of minerals. The goals of mineralogical studies may be quite diverse, ranging from the description and classification of a new or rare mineral, to an analysis of crystal structure involving determination of its internal atomic arrangement, or to the laboratory or industrial synthesis of mineral species at high temperatures and pressures. The methods employed in such studies are equally varied and include simple physical and chemical identification tests, determination of crystal symmetry, optical examination, X-ray diffraction, isotopic analysis, and other sophisticated procedures.

Although much mineralogical research centres on the chemical and physical properties of minerals, significant work is conducted on their origin as well. Investigators are frequently able to infer the way in which a mineral species forms on the basis of data obtained from laboratory experiments and on theoretical principles drawn from physical chemistry and thermodynamics.

Optical Mineralogy

Optical mineralogy is the study of minerals and rocks by measuring their optical properties. Most commonly, rock and mineral samples are prepared as thin sections or grain mounts for study in the laboratory with a petrographic microscope. Optical mineralogy is used to identify the mineralogical composition of geological materials in order to help reveal their origin and evolution.

Some of the properties and techniques used include:

- Refractive index

- Birefringence

- Michel-Lévy Interference colour chart

- Pleochroism

- Extinction angle

- Conoscopic interference pattern (Interference figure)

- Becke line test

- Optical relief

- Sign of elongation (Length fast vs. length slow)

- Wave plate

A rock-section should be about one-thousandth of an inch (30 micrometres) in thickness, and is relatively easy to make. A thin splinter of the rock, about 1 centimetre may be taken; it should be as fresh as possible and free from obvious cracks. By grinding it on a plate of planed steel or cast iron with a little fine carborundum it is soon rendered flat on one side, and is then transferred to a sheet of plate glass and smoothed with the finest grained emery until all roughness and pits are removed, and the surface is a uniform plane. The rock chip is then washed, and placed on a copper or iron plate which is heated by a spirit or gas lamp. A microscopic glass slip is also warmed on this plate with a drop of viscous natural Canada balsam on its surface. The more volatile ingredients of the balsam are dispelled by the heat, and when that is accomplished the smooth, dry, warm rock is pressed firmly into contact with the glass plate so that the film of balsam intervening may be as thin as possible and free from air bubbles. The preparation is allowed to cool, and the rock chip is again ground down as before, first with carborundum and, when it becomes transparent, with fine emery until the desired thickness is obtained. It is then cleaned, again heated with an additional small amount of balsam, and covered with a cover glass. The labor of grinding the first surface may be avoided by cutting off a smooth slice with an iron disk armed with crushed diamond powder. A second application of the slitter after the first face is smoothed and cemented to the glass will, in expert hands, leave a section of rock so thin as to be transparent. In this way the preparation of a section may require only twenty minutes.

Microscope

The microscope employed is usually one which is provided with a rotating stage beneath which there is a polarizer, while above the objective or eyepiece an analyzer is mounted; alternatively the stage may be fixed, and the polarizing and analyzing prisms may be capable of simultaneous rotation by means of toothed wheels and a connecting rod. If ordinary light and not polarized light is desired, both prisms may be withdrawn from the axis of the instrument; if the polarizer only is inserted the light transmitted is plane polarized; with both prisms in position the slide is viewed in cross-polarized light, also known as "crossed nicols." A microscopic rock-section in ordinary light, if a suitable magnification (e.g. around 30x) be employed, is seen to consist of grains or crystals varying in color, size, and shape.

Photomicrograph of a volcanic lithic fragment (sand grain); upper picture is plane-polarized light, bottom picture is cross-polarized light, scale box at left-center is 0.25 millimeter.

Characteristics of Minerals

Some minerals are colorless and transparent (quartz, calcite, feldspar, muscovite, etc.), while others are yellow or brown (rutile, tourmaline, biotite), green (diopside, hornblende, chlorite), blue (glaucophane), pink (garnet), etc. The same mineral may present a variety of colors, in the same or different rocks, and these colors may be arranged in zones parallel to the surfaces of the crystals. Thus tourmaline may be brown, yellow, pink, blue, green, violet, grey, or colorless, but every mineral has one or more characteristic, most common tints. The shapes of the crystals determine in a general way the outlines of the sections of them presented on the slides. If the mineral has one or more good cleavages, they will be indicated by systems of cracks. The refractive index is also clearly shown by the appearance of the section, which are rough, with well-defined borders if they have a much stronger refraction than the medium in which they are mounted. Some minerals decompose readily and become turbid and semi-transparent (e.g. feldspar); others remain always perfectly fresh and clear (e.g. quartz), while others yield characteristic secondary products (such as green chlorite after biotite). The inclusions in the crystals (both solid and fluid) are of great interest; one mineral may enclose another, or may contain spaces occupied by glass, by fluids or by gases.

Microstructure

The structure of the rock - the relation of its components to one another - is usually clearly indicated, whether it is fragmented or massive; the presence of glassy matter in contradistinction to a completely crystalline or "holo-crystalline" condition; the nature and origin of organic fragments; banding, foliation or lamination; the pumiceous or porous structure of many lavas. These and many other characters, though often not visible in the hand specimens of a rock, are rendered obvious by the examination of a microscopic section. Various methods of detailed observation may be applied, such as the measurement of the size of the elements of the rock by the help of micrometers, their relative proportions by means of a glass plate ruled in small squares, the angles between cleavages or faces seen in section by the use of the rotating graduated stage, and the estimation of the refractive index of the mineral by comparison with those of different mounting media.

Pleochroism

The light vibrates in only one plane, and in passing through doubly refracting crystals in the slide, is, speaking generally, broken up into rays, which vibrate at right angles to one another. In many colored minerals such as biotite, hornblende, tourmaline, chlorite, these two rays have different colors, and when a section containing any of these minerals is rotated the change of color is often clearly noticeable. This property, known as "pleochroism" is of great value in the determination of mineral composition.

Pleochroism is often especially intense in small spots which surround minute enclosures of other minerals, such as zircon and epidote. These are known as "pleochroic halos."

Double Refraction

If the analyzer is inserted in such a position that it is crossed relatively to the polarizer, the field of view will be dark where there are no minerals or where the light passes through isotropic substances such as glass, liquids and cubic crystals. All other crystalline bodies, being doubly refracting, will appear bright in some position as the stage is rotated. The only exception to this rule is provided by sections which are perpendicular to the optic axes of birefringent crystals, which remain dark or nearly dark during a whole rotation, the investigation of which is frequently important.

Extinction

Doubly refracting mineral sections will in all cases appear black in certain positions as the stage is rotated. They are said to "go extinct" when this takes place. The angle between these and any cleavages can be measured by rotating the stage and recording these positions. These angles are characteristic of the system to which the mineral belongs, and often of the mineral species itself. To facilitate measurement of extinction angles, various types of eyepieces have been devised, some having a stereoscopic calcite plate, others with two or four plates of quartz cemented together. These are often found to give more precise results than are obtained by observing only the position in which the mineral section is most completely dark between crossed nicols.

The mineral sections when not extinguished are not only bright, but are colored, and the colors they show depend on several factors, the most important of which is the strength of the double refraction. If all the sections are of the same thickness, as is nearly true of well-made slides, the minerals with strongest double refraction yield the highest polarization colors. The order in which the colors are arranged is expressed in what is known as Newton's scale, the lowest being dark grey, then grey, white, yellow, orange, red, purple, blue, and so on. The difference between the refractive indexes of the ordinary and the extraordinary ray in quartz is 0.009, and in a rock-section about 1/500 of an inch thick, this mineral gives grey and white polarization colors; nepheline with weaker double refraction gives dark grey; augite on the other hand will give red and blue, while calcite with the stronger double refraction will appear pinkish or greenish white. All sections of the same mineral, however, will not have the same color: sections perpendicular to an optic axis will be nearly black, and, in general, the more nearly any section approaches this direction the lower its polarization colors will be. By taking the average, or the highest color given by any mineral, the relative value of its double refraction can be estimated, or if the thickness of the section be precisely known the difference between the two refractive indexes can be ascertained. If the slides are thick the colors will be on the whole higher than on thin slides.

It is often important to find out whether of the two axes of elasticity (or vibration traces) in the section is that of greater elasticity (or lesser refractive index). The quartz wedge or selenite plate enables this. Suppose a doubly refracting mineral section so placed that it is extinguished; if now is rotated through 45 degrees it will be brightly illuminated. If the quartz wedge be passed across it so that the long axis of the wedge is parallel to the axis of elasticity in the section the polarization colors will rise or fall. If they rise the axes of greater elasticity in the two minerals are parallel; if they sink the axis of greater elasticity in the one is parallel to that of lesser elasticity in the other. In the latter case by pushing the wedge sufficiently far complete darkness or compensation will result. Selenite wedges, selenite plates, mica wedges and mica plates are also used for this purpose. A quartz wedge also may be calibrated by determining the amount of double refraction in all parts of its length. If now it be used to produce compensation or complete extinction in any doubly refracting mineral section, we can ascertain what is the strength of the double refraction of the section because it is obviously equal and opposite to that of a known part of the quartz wedge.

A further refinement of microscopic methods consists of the use of strongly convergent polarized light (conoscopic methods). This is obtained by a wide angled achromatic condenser above the polarizer, and a high power microscopic objective. Those sections are most useful which are perpendicular to an optic axis, and consequently remain dark on rotation. If they belong to uniaxial crystals they show a dark cross or convergent light between crossed nicols, the bars of which remain parallel to the wires in the field of the eyepiece. Sections perpendicular to an optic axis of a biaxial mineral under the same conditions show a dark bar which on rotation becomes curved to a hyperbolic shape. If the section is perpendicular to a "bisectrix" a black cross is seen which on rotation opens out to form two hyperbolas, the apices of which are turned towards one another. The optic axes emerge at the apices of the hyperbolas and may be surrounded by colored rings, though owing to the thinness of minerals in rock sections these are only seen when the double refraction of the mineral is strong. The distance between the axes as seen in the field of the microscope depends partly on the axial angle of the crystal and partly on the numerical aperture of the objective. If it is measured by means of eye-piece micrometer, the optic axial angle of the mineral can be found by a simple calculation. The quartz wedge, quarter mica plate or selenite plate permit the determination of the positive or negative character of the crystal by the changes in the color or shape of the figures observed in the field. These operations are similar to those employed by the mineralogist in the examination of plates cut from crystals.

Examination of Rock Powders

Although rocks are now studied principally in microscopic sections the investigation of fine crushed rock powders, which was the first branch of microscopic petrology to receive attention, is still actively used. The modern optical methods are readily applicable to transparent mineral fragments of any kind. Minerals are almost as easily determined in powder as in section, but it is otherwise with rocks, as the structure or relation of the components to one another. This is an element of great importance in the study of the history and classification of rocks, and is almost completely destroyed by grinding them to powder.

Magnetic Mineralogy

Magnetic mineralogy is the study of the magnetic properties of minerals. The contribution of a mineral to the total magnetism of a rock depends strongly on the type of magnetic order or disorder. Magnetically disordered minerals (diamagnets and paramagnets) contribute a weak magnetism and have no remanence. The more important minerals for rock magnetism are the minerals that can be magnetically ordered, at least at some temperatures. These are the ferromagnets, ferrimagnets and certain kinds of antiferromagnets. These minerals have a much stronger response to the field and can have a remanence.

Weakly Magnetic Minerals

Non-Iron-bearing Minerals

Most minerals with no iron content are diamagnetic. Some such minerals may have a significant positive magnetic susceptibility, for example serpentine, but this is because the minerals have inclusions containing strongly magnetic minerals such as magnetite. The susceptibility of such minerals is negative and small.

Mineral	Volume susceptibility at room temperature $\times$ 10^{-6} (SI)
Graphite	-80 to -200
Calcite	-7.5 to -39
Anhydrite	-14 to -60
Gypsum	-13 to -29
Ice	-9
Orthoclase	-13 to -17
Magnesite	-15
Forsterite	-12
Halite	-10 to -16
Galena	-33
Quartz	-13 to -17
Celestine	-16 to -18
Sphalerite	-31 to -750

Iron-bearing Paramagnetic Minerals

Reddish crystals: Biotite.

Most iron-bearing carbonates and silicates are paramagnetic at all temperatures. Some sulfides are paramagnetic, but some are strongly magnetic. In addition, many of the strongly magnetic minerals discussed below are paramagnetic above a critical temperature (the Curie temperature or Néel temperature). In table below are given susceptibilities for some iron-bearing minerals. The susceptibilities are positive and an order of magnitude or more larger than diamagnetic susceptibilities.

Table: Susceptibilities of some paramagnetic minerals.

Mineral	Volume susceptibility $\times 10^{-6}$ (SI)
Garnet	2,700
Illite	410
Montmorillonite	330-350
Biotite	1,500-2,900
Siderite	1,300-11,000
Chromite	3,000-120,000
Orthopyroxene	1,500-1,800
Fayalite	5,500
Olivine	1,600
Jacobsite	25,000
Franklinite	450,000

Strongly Magnetic Minerals

Iron-Titanium Oxides

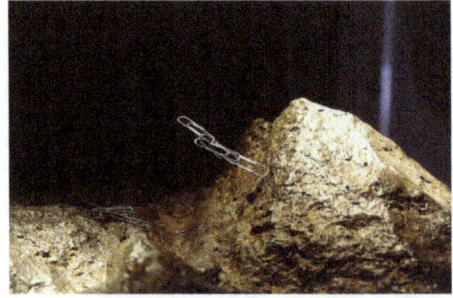

Magnetite-bearing lodestone displaying strong magnetic properties.

Many of the most important magnetic minerals on Earth are oxides of iron and titanium. Their compositions are conveniently represented on a ternary plot with axes corresponding to the proportions of Ti^{4+}, Fe^{2+}, and Fe^{3+}. Important regions on the diagram include the *titanomagnetites*, which form a line of compositions $Fe_{3-x}Ti_xO_4$ for x between 0 and 1. At the $x=0$ end is magnetite, while the $x=1$ composition is ulvöspinel. The titanomagnetites have an inverse spinel crystal structure and at high temperatures are a solid solution series. Crystals formed from titanomagnetites by cation-deficient oxidation are called *titanomaghemites*, an important example of which is maghemite. Another series, the *titanohematites*, have hematite and ilmenite as their end members, and so are also called *hemoilmenites*. The crystal structure of hematite is trigonal-hexagonal. It has the same composition as maghemite; to distinguish between them, their chemical formulae are generally given as γFe_2O_3 for hematite and αFe_2O_3 for maghemite.

Iron Sulfides

The other important class of strongly magnetic minerals is the iron sulfides, particularly greigite and pyrrhotite.

Iron Alloys

Meteorite slice with intergrowth of kamacite and taenite.

Extraterrestrial environments being low in oxygen, minerals tend to have very little Fe^{3+}. The primary magnetic phase on the Moon is ferrite, the body-centered cubic (bcc) phase of iron. As the proportion of iron decreases, the crystal structure changes from bcc to face centered cubic (fcc). Nickel iron mixtures tend to exsolve into a mixture of iron-rich kamacite and iron-poor taenite.

Environmental Mineralogy

Environmental mineralogy deals with the minerals taking part in constructing basic environmental systems. Minerals sometimes play crucial role in environmental degradation. Rock layers excavated for mining, mine and industrial waste dumps, abandoned mines etc. are sources of minerals that may lead to environmental pollution which is a matter of serious concern nowadays. Environmental mineralogy, in an interdisciplinary approach, deals with the interaction of the minerals with the biogeochemical environment. The objective of the study of environmental mineralogy is to provide detailed information regarding mineral-related environmental problems like effects of minerals on human health, formation of surfacial acidic environments, microbemineral interactions etc. The scope of environmental mineralogical study is to carry out mineralogical research that ultimately contributes to the development of effective solutions to these problems. The primary parameters, needed to determine these minerals, are identification of minerals, determination of quantity of the phases, together with compositional and textural information, and also microstructure or mineral surface chemistry. Apart from natural minerals, synthetic minerals are also taken into consideration to study their effects on environmental mineralogy. To understand and clarify the basic principles of environmental mineralogy, some aspects are needed to be discussed in details, like, various types of pollutions, health hazards from natural minerals, infrastructures of mineral industry and environmental factors and mining and mineral industry.

Clay Mineralogy

Clay minerals are a group of hydrous aluminium phyllosilicates, characterised by two-dimensional sheet structures. Variable cation substitution in these sheets leads to differential layer charge which result in changeable reactions with water and organics together with high surface area and cation exchange capacities.

Clay minerals can be divided into four major groups: kaolin, smectite, illite and chlorite. Each of these groups exhibit different characteristics. Clay minerals are typically fine-grained (4 μm) and constitute about 16% volume of material at the Earth's surface and are abundant in soils, sedimentary and low-grade metamorphic rocks and hydrothermal alteration zones.

Importance of Clay Minerals

Studies of clay minerals are imperative for a wide range of stakeholders, for example:

- Soils and agronomy – Clay minerals are fundamental to many soil functions including water, nutrient (e.g. potassium, ammonium) and contaminant (pesticides, heavy metals) retention, carbon storage, maintaining soil structure and the filtering of both ground and surface waters.

- Civil engineering including waste disposal – Due to their large shrink-swell and sorptive capacities, studies of clay minerals are essential to site assessments and geotechnical investigations. Clay minerals are also frequently employed as engineered materials.

- Energy – Studies of clay minerals may be employed to detect the depositional environment, stratigraphic correlation, reservoir quality, cap rock properties and basin history of hydrocarbon deposits and for carbon capture and storage.

- Industry – Clay minerals form critical raw materials for a range of industrial processes including catalysis, colloids, paper coating, drilling fluids, pharmaceuticals, ceramics and nanocomposites.

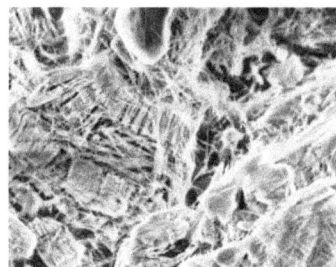

Halloysite.

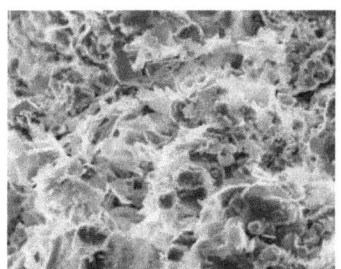

Saponite.

Clay mineralogy is the scientific discipline concerned with all aspects of clay minerals, including their properties, composition, classification, crystal structure, and occurrence and distribution in nature. The methods of study include X-ray diffraction, infrared spectroscopic analysis, chemical analyses of bulk and monomineralic samples, determinations of cationic exchange capacities, electron-optical studies, thermal studies by differential thermal analysis, and thermo-gravimetric methods. The singular character of these hydrous silicate minerals and the nature of the problems surrounding them justify the status of clay mineralogy as a discipline distinct from mineralogy. A common body of knowledge is involved, however, particularly with regard to studies of internal structure.

Referencess

- Mineralogy, science: britannica.com, Retrieved 21 May, 2019

- Nelson, Stephen A. "Interference Phenomena, Compensation, and Optic Sign". EENS 2110: Mineralogy. Tulane University. Retrieved 24 March 2017

- Clay, mpb, laboratories, sciencefacilities: bgs.ac.uk, Retrieved 22 June, 2019

- Clay-mineralogy, science: britannica.com, Retrieved 23 July, 2019

Mineral Analysis and Identification

- **Ways to Identify Common Minerals**
- **Mineral Identification**
- **Biomineralization**

Some of the ways of identifying a mineral are by analyzing how it reflects light, testing its hardness, identify fractures, etc. A deeper study into the analysis of minerals can be conducted using tools such as electron microscopes. The topics elaborated in this chapter will help in gaining a better perspective about the ways of identifying minerals and analyzing them.

Ways to Identify Common Minerals

Since minerals are the building blocks of rocks, it is important that you learn to identify the most common varieties. Minerals can be distinguished using various physical and/or chemical characteristics, but, since chemistry cannot be determined readily in the field, geologists use the physical properties of minerals to identify them.

These include features such as crystal form, hardness (relative to a steel blade or you finger nail), colour, lustre, and streak (the colour when a mineral is ground to a powder). Generally the characteristics listed above can only be determined if the mineral grains are visible in a rock. Thus the identification key distinguishes between rocks in which the grains are visible and those in which the individual mineral components are too small to identify.

The descriptions below use various terms or numbers to describe the mineral's shape, hardness, appearance after breaking, or other attributes.

Crystalline minerals are most often quartz. Quartz is an extremely common mineral, and its glittering or crystalline appearance catches the eye of many collectors. Quartz has a hardness of 7 on the Mohs scale, and demonstrates any type of fracture when broken, never the flat surface of cleavage. It does not leave a noticeable streak on white porcelain. It has a glassy luster, or shine.

Quartz cluster

- Milky quartz is white quartz. The white color comes from carbon dioxide gas trapped within the quartz structure. Milky quartz is usually massive, but well well-formed crystals are also common. In the Huachuca mountains, milky quartz occurs a a filling material in fractures (mineral veins). All quartz has a hardness of 7 on Mohs scale of hardness and can easily scratch glass. Milky quartz is shiny and translucent. Quartz has no cleavage and breaks with a fracture that ranges from conchoidal to irregular.

- Rose quartz is a variety of massive, translucent quartz with a pink color. It has no cleavage, it breaks with a conchoidal fracture, and it has a shiny surface. Depending on quality, it can be used as a gemstone or a decorative garden stone. Two major occurrences of rose quartz are Maine and the Black Hills of South Dakota.

- Amethyst is purple quartz. It can occur as well-formed quartz crystals in geodes or deformed crystals in a mineral vein. If the quality is high enough, amethyst is used as a gemstone. The luster of amethyst is usually vitreous (shiny).

Chert

Chert with flint

All types of quartz are crystalline, but some varieties, called "cryptocrystalline," are made of minuscule crystals not visible to the eye.If the mineral has a hardness of 7, fractures, and has a glassy luster, it may be a type of quartz called chert. This is most commonly brown or grey.

"Flint" is one variety of chert, but it is categorized in many different ways. For instance, some people may refer to any black chert as flint, while others may only call it flint if it has a certain luster or was found among certain types of rock.

Chalcedony

Chalcedony is formed from a mixtures of quartz and another mineral, moganite. There are many beautiful varieties, typically forming striped bands of different colors. Here are two of the most common:

- Onyx is a type of chalcedony that tends to have parallel bands. It is most often black or white, but can be many colors.

- Agate has more curving or "wiggly" bands, and can show up in a wide variety of different colors. It can form from pure quartz, chalcedony, or similar minerals.

Feldspar

Orthoclase, Plagioclase and Amazonite

Besides the many varieties of quartz, feldspar is the most common type of mineral found. Feldspar is the other common, light-coloured rock-forming mineral. Instead of being glassy like quartz, it is generally dull to opaque with a porcelain-like appearance. Colour varies from red, pink, and white (orthoclase) to green, grey and white (plagioclase). Feldspar is also hard but can be scratched by quartz. Feldspar in igneous rocks forms well developed crystals which are roughly rectangular in shape, and they cleave or break along flat faces. The grains, in contrast to quartz, often have straight edges and flat rectangular faces, some of which meet at right angles.

Mica

Muscovite and Biotite

Mica is easily distinguished by its characteristic of peeling into many thin flat smooth sheets or flakes. This is similar to the cleavage in feldspar except that in the case of mica the cleavage planes are in only one direction and no right angle face joins occur. Mica may be white and pearly (muscovite) or dark and shiny (biotite).

- Muscovite mica is colorless to a very pale brown in color. It peels easily into very thin, flexible, elastic sheets that are nearly colorless. Muscovite is also known as white mica.

- Biotite mica ranges from dark brown to black. It also peels in very thin, flexible, elastic sheets like muscovite mica. Biotite is also known as black mica.

Difference between Gold and Fool's Gold: Pyrite

Pyrite is also known as "fool's gold" because it has a yellow metallic color. Pyrite can be distinguished from native gold by several different properties. Pyrite is much harder than gold; it cannot be scratched by a steel straight pin. Pyrite is brittle; it can be crushed to a powder, whereas gold simply flattens out because it is a metal. A streak test can also distinguish pyrite from gold; pyrite produces a greenish black streak and gold produces a yellow streak.

- Marcasite is another common mineral similar to pyrite. While pyrite crystals are shaped like cubes, marcasite forms needles.

Green and Blue Minerals are Often Malachite or Azurite

Both of these minerals contain copper, among other minerals. The copper gives malachite its rich green color, while it causes azurite to appear bright blue. These often occur together, and both have a hardness between 3 and 4.

Stunning piece of azurite and malachite from China.

- Azurite is a bright blue mineral associated with copper ore. It may occur with green malachite, also a copper ore. It is relatively soft at 3.5 on Mohs scale of hardness.

- Malachite is a rich green to dark green copper mineral. It can occur on its own or with azurite, a mineral that it is closely related to in chemistry. It is relatively soft at 3.5 on Mohs scale of hardness.

Gypsum

Gypsum from Annabel Lee mine, Hardin Co., Illinois, United States.

Gypsum is a soft, light-colored mineral. Its color can be colorless and transparent (selenite) or white, pale pink or pale brown. If crystallized, it displays one direction of excellent cleavage, but the cleavage fragments are much thicker than those of mica and the fragments are not elastic. Generally, it lacks the greasy feel of talc. One form of gypsum tends to form with a fibrous structure (satinspar).

Calcite

Variety of colors and forms of calcite.

Calcite is a very common mineral. The difficulty in identifying it is that can occur in a very large variety of colors and forms. One of the most common forms of calcite crystals are pointy pyramids that resemble a dog's canine tooth (dogtoothspar). Large, pure pieces of crystalline calcite display three directions of cleavage that are inclined (not at 90 degrees).

Calcite ranges from transparent to translucent. Colors may be colorless, white, cream, pale yellow, yellow-brown, brown, and even red due to impurities. The easiest way to distinguish calcite is with an acid test; concentrated hydrochloric acid with cause abundant bubbles to form as it reacts with the calcite.

Fluorite

Fluorite-Octahedron

It is sometimes easy to mistake fluorite for calcite on a quick examination. However, if you pay careful attention, fluorite has four directions of cleavage compared to three directions of cleavage for calcite. Fluorite is also harder than calcite (4 on Mohs scale of hardness) and can scratch a piece of calcite. Fluorite is often more colorful than calcite and can be purple, green, yellow, pink, brown, or colorless and may even show two or more colors on the same specimen. Fluorite crystals are usually cubes or octahedrons. Above all, fluorite does not fizz in contact with hydrochloric acid.

Magnetite

Perfect octahedron Magnetite from Formazza Valley.

Magnetite is common in igneous and metamorphic rocks, and some sediments, though usually in only small amounts (1 - 2 %). It is black in colour with a metallic lustre, occurring in small octahedra (like two pyramids stuck together). Easily recognized by its strongly magnetic character.

Mineral Identification

Minerals can be identified by their physical characteristics. The physical properties of minerals are related to their chemical composition and bonding. Some characteristics, such as a mineral's hardness, are more useful for mineral identification. Color is readily observable and certainly obvious, but it is usually less reliable than other physical properties.

How are Minerals Identified?

Mineralogists are scientists who study minerals. One of the things mineralogists must do is identify and categorize minerals. While a mineralogist might use a high-powered microscope to identify some minerals, most are recognizable using physical properties.

This mineral has shiny, gold, cubic crystals with striations, so it is pyrite.

Color, Streak and Luster

Diamonds are popular gemstones because the way they reflect light makes them very sparkly. Turquoise is prized for its striking greenish-blue color. Notice that specific terms are being used to describe the appearance of minerals.

Color

Color is rarely very useful for identifying a mineral. Different minerals may be the same color. Real gold, as seen in figure below, is very similar in color to the pyrite in figure above.

This mineral is shiny, very soft, heavy, and gold in color, and is actually gold.

The same mineral may also be found in different colors. Figure below shows one sample of quartz that is colorless and another quartz that is purple. A tiny amount of iron makes the quartz purple. Many minerals are colored by chemical impurities.

Purple quartz, known as amethyst, and clear quartz are the same mineral despite the different colors.

Streak

Streak is the color of a mineral's powder. Streak is a more reliable property than color because streak does not vary. Minerals that are the same color may have a different colored streak. Many minerals, such as the quartz in the figure above, do not have streak.

To check streak, scrape the mineral across an unglazed porcelain plate (figure below). Yellow-gold pyrite has a blackish streak, another indicator that pyrite is not gold, which has a golden yellow streak.

The streak of hematite across an unglazed porcelain plate is red-brown.

Luster

Luster describes the reflection of light off a mineral's surface. Mineralogists have special terms to describe luster. One simple way to classify luster is based on whether the mineral is metallic or non-metallic. Minerals that are opaque and shiny, such as pyrite, have a metallic luster. Minerals such as quartz have a non-metallic luster. Different types of non-metallic luster are described in table below.

Table: Six Types of Non-metallic Luster.

Luster	Appearance
Adamantine	Sparkly
Earthy	Dull, clay-like
Pearly	Pearl-like
Resinous	Like resins, such as tree sap
Silky	Soft-looking with long fibers
Vitreous	Glassy

Hardness

Hardness is a measure of whether a mineral will scratch or be scratched. Mohs Hardness Scale, shown in Table below, is a reference for mineral hardness.

Table: Mohs Hardness Scale - 1 (Softest) to 10 (Hardest).

Hardness	Mineral
1	Talc
2	Gypsum
3	Calcite
4	Fluorite
5	Apatite
6	Feldspar
7	Quartz
8	Topaz
9	Corundum
10	Diamond

With a Mohs scale, anyone can test an unknown mineral for its hardness. Imagine you have an unknown mineral. You find that it can scratch fluorite or even apatite, but feldspar scratches it. You know then that the mineral's hardness is between 5 and 6. Note that no other mineral can scratch diamond.

Cleavage and Fracture

Breaking a mineral breaks its chemical bonds. Since some bonds are weaker than other bonds, each type of mineral is likely to break where the bonds between the atoms are weaker. For that reason, minerals break apart in characteristic ways.

Cleavage is the tendency of a mineral to break along certain planes to make smooth surfaces. Halite breaks between layers of sodium and chlorine to form cubes with smooth surfaces (figure below).

A close-up view of sodium chloride in a water bubble aboard the International Space Station.

Mica has cleavage in one direction and forms sheets (figure below).

Sheets of mica.

Minerals can cleave into polygons. Fluorite forms octahedrons (Figure below).

This rough diamond shows its octahedral cleavage.

One reason gemstones are beautiful is that the cleavage planes make an attractive crystal shape with smooth faces.

Fracture is a break in a mineral that is not along a cleavage plane. Fracture is not always the same in the same mineral because fracture is not determined by the structure of the mineral. Minerals may have characteristic fractures (figure below). Metals usually fracture into jagged edges. If a mineral splinters like wood, it may be fibrous. Some minerals, such as quartz, form smooth curved surfaces when they fracture.

Chrysotile has splintery fracture.

Other Identifying Characteristics

Some minerals have other unique properties.

Table: Some minerals have unusual properties that can be used for identification.

Property	Description	Example of Mineral
Fluorescence	Mineral glows under ultraviolet light.	Fluorite
Magnetism	Mineral is attracted to a magnet.	Magnetite
Radioactivity	Mineral gives off radiation that can be measured with Geiger counter.	Uraninite
Reactivity	Bubbles form when mineral is exposed to a weak acid.	Calcite
Smell	Some minerals have a distinctive smell.	Sulfur (smells like rotten eggs)
Taste	Some minerals taste salty.	Halite

Electron Microscopy of Minerals

Scanning electron microscopy is now a routine technology employed in the study of rocks and minerals. In addition to providing high-resolution images, electron microscopes generate a variety of additional signals that are often employed to further our understanding of geological samples. Chief among these are back-scatter electron detection, secondary electron detection, and energy-dispersive X-ray spectroscopy.

- The intensity of back-scattered electrons (BSEs) is directly proportional to the mean atomic number of the sample. For this reason, BSE detection allows researchers to visualize different phases and grains directly in polished samples. Contrast can be set to distinguish atomic number differences as close as 0.1z. BSE detection further allows for the automated measurement of grain size distributions, nearest neighbor distances, and phase coverage by area.

- Secondary electrons (SEs) arise from the absolute surface of the material and are generally used to study the morphology of 3-dimensional materials. High surface sensitivity also results in enhanced resolution and improved depth of field when compared to a BSE detector.

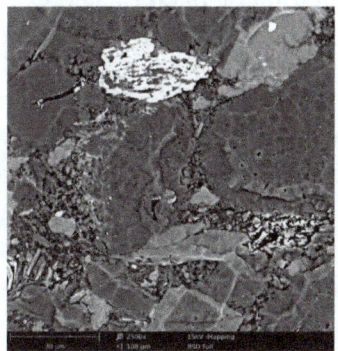 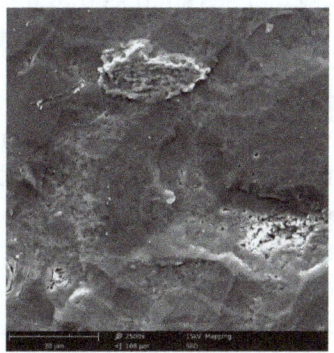

BSE Image of a Petrographic Thin Section. BSE image of the same Petrographic Thin Section.

After different regions of interest have been identified, energy dispersive X-ray spectroscopy can be used to quantify the elemental composition of these phases. At each pixel in the image below, a full X-ray spectrum is acquired, then the pixel is colored according to the highest intensity element.

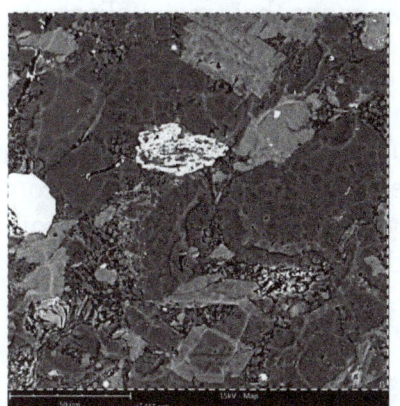 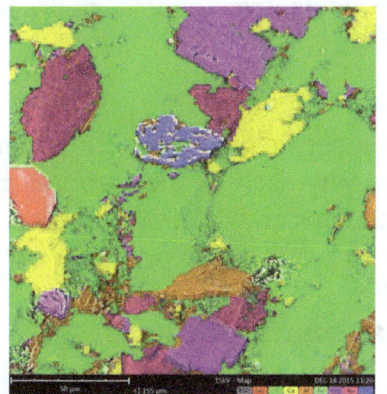

BSE image of the Petrographic Thin Section. Corresponding EDS Map analysis of the same Petrographic Thin Section.

The Phenom SEM also incorporates an integrated motorized X/Y stage that can be used to acquire and stitch high-resolution mosaics over large regions. These images can be further processed to measure grain size distributions, nearest neighbor distances, and area coverage.

Biomineralization

Biomineralization is the process by which living organisms produce minerals, often to harden or stiffen existing tissues. Such tissues are called mineralized tissues. It is an extremely widespread phenomenon;

all six taxonomic kingdoms contain members that are able to form minerals, and over 60 different minerals have been identified in organisms. Examples include silicates in algae and diatoms, carbonates in invertebrates, and calcium phosphates and carbonates in vertebrates. These minerals often form structural features such as sea shells and the bone in mammals and birds. Organisms have been producing mineralised skeletons for the past 550 million years. Ca carbonates and Ca phosphates are usually crystalline, but silica organisms (sponges, diatoms, etc.) are always non crystalline minerals. Other examples include copper, iron and gold deposits involving bacteria. Biologically-formed minerals often have special uses such as magnetic sensors in magnetotactic bacteria (Fe_3O_4), gravity sensing devices ($CaCO_3$, $CaSO_4$, $BaSO_4$) and iron storage and mobilization ($Fe_2O_3.H_2O$ in the protein ferritin).

In terms of taxonomic distribution, the most common biominerals are the phosphate and carbonate salts of calcium that are used in conjunction with organic polymers such as collagen and chitin to give structural support to bones and shells. The structures of these biocomposite materials are highly controlled from the nanometer to the macroscopic level, resulting in complex architectures that provide multifunctional properties. Because this range of control over mineral growth is desirable for materials engineering applications, there is significant interest in understanding and elucidating the mechanisms of biologically controlled biomineralization.

Biological Roles

Among metazoans, biominerals composed of calcium carbonate, calcium phosphate or silica perform a variety of roles such as support, defense and feeding. It is less clear what purpose biominerals serve in bacteria. One hypothesis is that cells create them to avoid entombment by their own metabolic byproducts. Iron oxide particles may also enhance their metabolism.

Biology

If present on a super-cellular scale, biominerals are usually deposited by a dedicated organ, which is often defined very early in the embryological development. This organ will contain an organic matrix that facilitates and directs the deposition of crystals. The matrix may be collagen, as in deuterostomes, or based on chitin or other polysaccharides, as in molluscs.

Shell Formation in Molluscs

Variety of mollusc shells (gastropods, snails and seashells).

The mollusc shell is a biogenic composite material that has been the subject of much interest in materials science because of its unusual properties and its model character for biomineralization. Molluscan shells consist of 95–99% calcium carbonate by weight, while an organic component makes up the remaining 1–5%. The resulting composite has a fracture toughness ≈3000 times greater than that of the crystals themselves. In the biomineralization of the mollusc shell, specialized proteins are responsible for directing crystal nucleation, phase, morphology, and growths dynamics and ultimately give the shell its remarkable mechanical strength. The application of biomimetic principles elucidated from mollusc shell assembly and structure may help in fabricating new composite materials with enhanced optical, electronic, or structural properties. The most described arrangement in mollusc shells is the nacre - prismatic shells, known in large shells as *Pinna* or the pearl oyster (*Pinctada*). Not only the structure of the layers differ, but their mineralogy and chemical composition also differ. Both contain organic components (proteins, sugars and lipids) and the organic components are characteristic of the layer, and of the species. The structures and arrangements of mollusc shells are diverse, but they share some features: the main part of the shell is a crystalline Ca carbonate (aragonite, calcite), despite some amorphous Ca carbonate occurs; and despite they react as crystals, they never show angles and facets. The examination of the inner structure of the prismatic units, nacreous tablets, foliated laths etc. shows irregular rounded granules.

Mineral Production and Degradation in Fungi

Fungi are a diverse group of organisms that belong to the eukaryotic domain. Studies of their significant roles in geological processes, "geomycology", has shown that fungi are involved with biomineralization, biodegradation, and metal-fungal interactions.

In studying fungi's roles in biomineralization, it has been found that fungi deposit minerals with the help of an organic matrix, such as a protein, that provides a nucleation site for the growth of biominerals. Fungal growth may produce a copper-containing mineral precipitate, such as copper carbonate produced from a mixture of $(NH_4)_2CO_3$ and $CuCl_2$. The production of the copper carbonate is produced in the presence of proteins made and secreted by the fungi. These fungal proteins that are found extracellularly aid in the size and morphology of the carbonate minerals precipitated by the fungi.

In addition to precipitating carbonate minerals, fungi can also precipitate uranium-containing phosphate biominerals in the presence of organic phosphorus that acts a substrate for the process. The fungi produce a hyphal matrix, also known as mycelium, that localizes and accumulates the uranium minerals that have been precipitated. Although uranium is often deemed as toxic towards living organisms, certain fungi such as *Aspergillus niger* and *Paecilomyces javanicus* can tolerate it.

Though minerals can be produced by fungi, they can also be degraded; mainly by oxalic-acid producing strains of fungi. Oxalic acid production is increased in the presence of glucose for three organic acid producing fungi – *Aspergillus niger*, *Serpula himantioides*, and *Trametes versicolor*. These fungi have been found to corrode apatite and galena minerals. Degradation of minerals by fungi is carried out through a process known as neogenesis. The order of most to least oxalic acid secreted by the fungi studied are *Aspergillus niger*, followed by *Serpula himantioides*, and finally *Trametes versicolor*. These capabilities of certain groups of fungi have a major impact on corrosion, a costly problem for many industries and the economy.

Chemistry

Because extracellulariron is strongly involved in inducing calcification, its control is essential in developing shells; the protein ferritin plays an important role in controlling the distribution of iron. The most common mineral present in biomineralization is hydroxyapatite (HA), which is a naturally occurring mineral form of calcium apatite with the formula $Ca_{10}(PO_4)_6(OH)_2$. Hydroxyapatite crystals are found in many biological materials including bones, fish scales, and cartilage. Each material has a mineral content which corresponds with the required mechanical properties, where increasing HA content typically leads to increased stiffness but reduced extensibility.

Evolution

Some calcareous sponges.

The first evidence of biomineralization dates to some 750 million years ago, and sponge-grade organisms may have formed calcite skeletons 630 million years ago. But in most lineages, biomineralization first occurred in the Cambrian or Ordovician periods. Organisms used whichever form of calcium carbonate was more stable in the water column at the point in time when they became biomineralized, and stuck with that form for the remainder of their biological history. The stability is dependent on the Ca/Mg ratio of seawater, which is thought to be controlled primarily by the rate of sea floor spreading, although atmospheric CO_2 levels may also play a role.

Biomineralization evolved multiple times, independently, and most animal lineages first expressed biomineralized components in the Cambrian period. Many of the same processes are used in unrelated lineages, which suggests that biomineralization machinery was assembled from pre-existing "off-the-shelf" components already used for other purposes in the organism. Although the biomachinery facilitating biomineralization is complex – involving signalling transmitters, inhibitors, and transcription factors – many elements of this 'toolkit' are shared between phyla as diverse as corals, molluscs, and vertebrates. The shared components tend to perform quite fundamental tasks, such as designating that cells will be used to create the minerals, whereas genes controlling more finely tuned aspects that occur later in the biomineralization process – such as the precise alignment and structure of the crystals produced – tend to be uniquely evolved in different lineages. This suggests that Precambrian organisms were employing the same elements, albeit for a different purpose — perhaps to *avoid* the inadvertent precipitation of calcium carbonate from the supersaturated Proterozoic oceans. Forms of mucus that are involved in inducing mineralization in most metazoan lineages appear to have performed such an anticalcifatory function in the ancestral state. Further, certain proteins that would originally have been involved in maintaining calcium concentrations within cellsare homologous to all metazoans, and appear to have been co-opted into biomineralization after the divergence of the metazoan lineages. The *galaxins* are one probable example of a gene being co-opted from a different ancestral purpose into controlling biomineralization, in this case being 'switched' to this purpose in the Triassic scleractinian corals; the role performed appears to be functionally identical to the unrelated pearlin gene in molluscs.Carbonic anhydrase serves a role in mineralization in sponges, as well as metazoans, implying an ancestral role. Far from being a rare trait that evolved a few times and remained stagnant, biomineralization pathways in fact evolved many times and are still evolving rapidly today; even within a single genus it is possible to detect great variation within a single gene family.

The homology of biomineralization pathways is underlined by a remarkable experiment whereby the nacreous layer of a molluscan shell was implanted into a human tooth, and rather than experiencing an immune response, the molluscan nacre was incorporated into the host bone matrix. This points to the exaptation of an original biomineralization pathway.

The most ancient example of biomineralization, dating back 2 billion years, is the deposition of magnetite, which is observed in some bacteria, as well as the teeth of chitons and the brains of vertebrates; it is possible that this pathway, which performed a magnetosensory role in the common ancestor of all bilaterians, was duplicated and modified in the Cambrian to form the basis for calcium-based biomineralization pathways. Iron is stored in close proximity to magnetite-coated chiton teeth, so that the teeth can be renewed as they wear. Not only is there a marked similarity between the magnetite deposition process and enamel deposition in vertebrates but some vertebrates even have comparable iron storage facilities near their teeth.

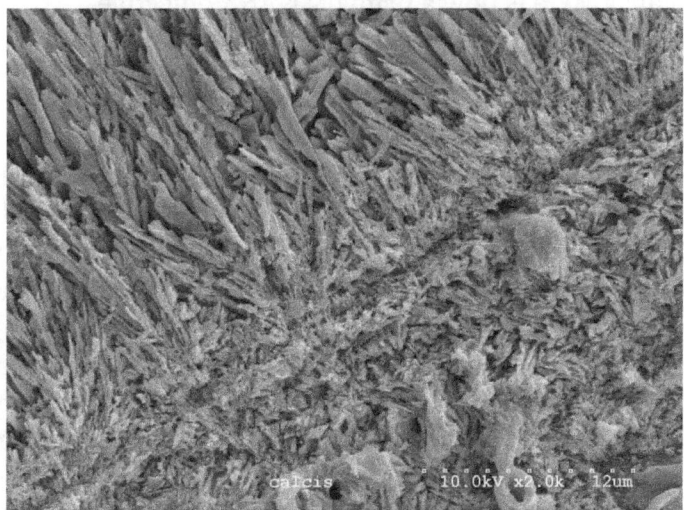

Glomerula piloseta, longitudinal section of the tube, aragonitic spherulitic prismatic structure

Type of mineralization	Examples of organisms
Calcium carbonate (calcite or aragonite)	• Foraminifera • Coccolithophores • Calcareous sponge spicules • Corals • Archaeocyatha • Bryozoans • Brachiopod and mollusc shells • Echinoderms • Serpulidae
Silica	• Radiolarians • Diatoms • Most sponge spicules
Apatite (phosphate carbonate)	• Enamel (vertebrate teeth) • Vertebrate bone • Conodonts

Astrobiology

It has been suggested that biominerals could be important indicators of extraterrestrial life and thus could play an important role in the search for past or present life on Mars. Furthermore, organic components (biosignatures) that are often associated with biominerals are believed to play crucial roles in both pre-biotic and biotic reactions.

On January 24, 2014, NASA reported that current studies by the *Curiosity* and *Opportunity* rovers on the planet Mars will now be searching for evidence of ancient life, including a biosphere based on autotrophic, chemotrophic and/or chemolithoautotrophic microorganisms, as well as ancient water, including fluvio-lacustrine environments (plains related to ancient rivers or lakes) that may have been habitable. The search for evidence of habitability, taphonomy (related to fossils), and organic carbon on the planet Mars is now a primary NASA objective.

Potential Applications

Most traditional approaches to synthesis of nanoscale materials are energy inefficient, requiring stringent conditions (e.g., high temperature, pressure or pH) and often produce toxic byproducts. Furthermore, the quantities produced are small, and the resultant material is usually irreproducible because of the difficulties in controlling agglomeration.In contrast, materials produced by organisms have properties that usually surpass those of analogous synthetically manufactured materials with similar phase composition. Biological materials are assembled in aqueous environments under mild conditions by using macromolecules. Organic macromolecules collect and transport raw materials and assemble these substrates and into short- and long-range ordered composites with consistency and uniformity. The aim of biomimetics is to mimic the natural way of producing minerals such as apatites. Many man-made crystals require elevated temperatures and strong chemical solutions, whereas the organisms have long been able to lay down elaborate mineral structures at ambient temperatures. Often, the mineral phases are not pure but are made as composites that entail an organic part, often protein, which takes part in and controls the biomineralisation. These composites are often not only as hard as the pure mineral but also tougher, as the micro-environment controls biomineralisation.

Uranium Contaminants in Groundwater

Biomineralization may be used to remediate groundwater contaminated with uranium. The biomineralization of uranium primarily involves the precipitation of uranium phosphate minerals associated with the release of phosphate by microorganisms. Negatively charged ligands at the surface of the cells attract the positively charged uranyl ion (UO_2^{2+}). If the concentrations of phosphate and UO_2^{2+} are sufficiently high, minerals such as autunite ($Ca(UO_2)_2(PO_4)_2.10\text{-}12H_2O$) or polycrystalline HUO_2PO_4 may form thus reducing the mobility of UO_2^{2+}. Compared to the direct addition of inorganic phosphate to contaminated groundwater, biomineralization has the advantage that the ligands produced by microbes will target uranium compounds more specifically rather than react actively with all aqueous metals. Stimulating bacterial phosphatase activity to liberate phosphate under controlled conditions limits the rate of bacterial hydrolysis of organophosphate and the release of phosphate to the system, thus avoiding clogging of the injection location with metal phosphate minerals. The high concentration of ligands near the cell surface also provides nucleation foci for precipitation, which leads to higher efficiency than chemical precipitation.

Autunite Crystal

Examples of biogenic minerals include:

- Apatite in bones and teeth.

- Aragonite, calcite, fluorite in vestibular systems (part of the inner ear) of vertebrates.

- Aragonite and calcite in travertine and biogenic silica (siliceous sinter, opal) deposited through algal action.

- Hydroxylapatite formed by mitochondria.

- Magnetite and greigite formed by magnetotactic bacteria.

- Pyrite and marcasite in sedimentary rocks deposited by sulfate-reducing bacteria.

- Quartz and diamonds formed from bacterial action on fossil fuels (gas, oil, coal).

- Goethite found as filaments in limpet teeth.

Bacterial calcium carbonate precipitation is a biomineralization process, which is a common phenomenon in the bacterial kingdom. It can be achieved by two different mechanisms, as either biologically-controlled or biologically-induced mineralization. In biologically-controlled mineralization, the organisms, such as magnetotatic bacteria, diatoms and coccolithophores, use specific metabolic and genetic pathways to control the process. However, calcium carbonate precipitation by bacteria is generally regarded as induced mineralization, as the types of minerals produced are dependent on the environmental conditions. This phenomenon occurs worldwide with numerous bacterial species, in various environments, such as soils, freshwaters, oceans and saline lakes, found to participate in the precipitation of mineral carbonates. These bacteria play a fundamental role in the calcium biogeochemical cycle, which contributes to the formation of calcium carbonate sediments, deposits and rocks.

Biologically-induced mineralization is usually carried out in open environments and the process is often linked to microbial cell surface structures and metabolic activities. Microbial extracellular polymeric substances (EPS) can trap and bind remarkable amounts of calcium to facilitate calcium carbonate precipitation, and most likely also play an essential role in calcium carbonate precipitation morphology and mineralogy. The mineralization process associated with microbial metabolic activities usually leads to an increase in environmental alkalinity, thereby facilitating calcium carbonate precipitation. Among these metabolic activities, the most common is urea hydrolysis catalyzed by urease enzymes, which commonly occurs in large varieties of microorganisms. The microbial urease enzyme hydrolyzes urea to produce carbonate and ammonia, increasing the pH and carbonate concentration, which then combines with environmental calcium to precipitate as calcium carbonate.

Calcite, aragonite and vaterite are three crystal polymorphs of calcium carbonate in bacterial systems, with calcite being the most common and stable bacterial carbonate polymorphs. Bacterial mineralization of aragonite, often representing the metastable polymorph, has also been reported. The production of the polymorphs of calcite, aragonite and vaterite depend both on their growing environments and bacterial strains. It was reported that different bacteria precipitated different types of calcium carbonate and were mainly either spherical or polyhedral crystalline forms. Bacterial-induced carbonate minerals have often been reported in a large number of bacteria, such as cyanobacteria, sulphate-reducing bacteria, Bacillus. Myxococcus Halobacteria and Pseudomonas. tested the crystal-producing ability among cave bacteria and found that all produced calcite except for Bacillus sp., which precipitated vaterites. reported that M. xanthus was able to induce precipitation of calcite and vaterite. Emerging evidence suggests that bacteria do not directly influence calcium carbonate morphology or polymorph selection. The morphological features instead may be influenced by the composition of the culture medium, the specific bacterial outer structures and their chemical nature, which might be crucial for the bacterial crystallization process.

References

- Harris, Ph.D., Edward D. (1 January 2014). Minerals in Food Nutrition, Metabolism, Bioactivity (1st ed.). Lancaster, PA: destech Publications, Inc. P. 378. ISBN 978-1-932078-97-8. Retrieved 30 January 2015

- How-to-identify-common-minerals: geologyin.com, Retrieved 24 August, 2019

- Astrid Sigel; Helmut Sigel; Rol, K.O. Sigel, eds. (2008). Biomineralization: From Nature to Application. Metal Ions in Life Sciences. 4. Wiley. ISBN 978-0-470-03525-2

- Mineral-identification, chapter, suny-earthscience: courses.lumenlearning.com, Retrieved 25 January, 2019

- Vinn, O. (2013). "Occurrence formation and function of organic sheets in the mineral tube structures of Serpulidae (Polychaeta Annelida)". Plos ONE. 8 (10): e75330. Bibcode:2013ploso...875330V. Doi:10.1371/journal.pone.0075330. PMC 3792063. PMID 24116035

- Electron-microscopy-minerals, geology, applications: nanoscience.com, Retrieved 27 March 2019

- Sherman, Vincent R. (2008). "The materials science of collagen". Journal of the Mechanical Behavior of Biomedical Materials. 61 (5): 529–534. Doi:10.1016/j.bjps.2007.06.004. PMID 17652049

Mineral Formation and Defects | 4

- **Mineral Formation**
- **Mineral Defects**
- **Structural Defects and Color in Minerals**
- **Mineral Stability and Phase Diagrams**
- **Formation of Minerals**
- **Crystal Twinning**

Minerals can be formed under a wide range of physical and chemical conditions such as fluctuating temperature and pressure. Defects in mineral crystals are the distortions of the ordered arrangement of atoms in the crystalline lattice. This chapter closely examines the key concepts related to the formation of minerals as well as the structural defects in them.

Mineral Formation

Minerals form within Earth or on Earth's surface by natural processes. Minerals develop when atoms of one or more elements join together and crystals begin to grow. Recall that each type of mineral has its own chemical makeup. Therefore, what types of minerals form in an area depends in part on which elements are present there. Temperature and pressure also affect which minerals form.

Water evaporates. Water usually has many substances dissolved in it. Minerals can form when the water evaporates. For example, when salt water evaporates, the atoms that make up halite, which is used as table salt, join to form crystals. Other minerals form from evaporation too, depending on the substances dissolved in the water. The mineral gypsum often forms as water evaporates.

Hot water cools. As hot water within Earth's crust moves through rocks, it can dissolve minerals. When the water cools, the dissolved minerals separate from the water and become solid again. In some cases, minerals are moved from one place to another. Gold can dissolve in hot water that moves through the crust. As the water cools and the gold becomes solid again, it can fill cracks in rocks. In other cases, the minerals that form are different from the ones that dissolved. Lead from the mineral galena can later become part of the mineral wulfenite as atoms join together into new minerals.

Molten rock cools. Many minerals grow from magma. Magma — molten rock inside Earth—contains all the types of atoms that are found in minerals. As magma cools, the atoms join together to form different minerals. Minerals also form as lava cools. is molten rock that has reached Earth's surface. Quartz is one of the many minerals that crystallize from magma and lava.

Heat and pressure cause changes. Heat and pressure within Earth cause new minerals to form as bonds between atoms break and join again. The mineral garnet can grow and replace the minerals chlorite and quartz as their atoms combine in new ways. The element carbon is present in some rocks. At high temperatures carbon forms the mineral graphite, which is used in pencils.

Organisms produce minerals. A few minerals are produced by living things. For example, ocean animals such as oysters and clams produce calcite and other carbonate minerals to form their shells. Even you produce minerals. Your body produces one of the main minerals in your bones and teeth—apatite.

In order for a mineral crystal to grow, the elements needed to make it must be present in the appropriate proportions, the physical and chemical conditions must be favourable, and there must be sufficient time for the atoms to become arranged.

Physical and chemical conditions include factors such as temperature, pressure, presence of water, pH, and amount of oxygen available. Time is one of the most important factors because it takes time for atoms to become ordered. If time is limited, the mineral grains will remain very small. The presence of water enhances the mobility of ions and can lead to the formation of larger crystals over shorter time periods.

Most of the minerals that make up the rocks around us formed through the cooling of molten rock, known as magma. At the high temperatures that exist deep within Earth, some geological materials are liquid. As magma rises up through the crust, either by volcanic eruption or by more gradual processes, it cools and minerals crystallize. If the cooling process is rapid (minutes, hours, days, or years), the components of the minerals will not have time to become ordered and only small crystals can form before the rock becomes solid. The resulting rock will be fine-grained (i.e., crystals less than 1 mm). If the cooling is slow (from decades to millions of years), the degree of ordering will be higher and relatively large crystals will form. In some cases, the cooling will be so fast (seconds) that the texture will be glassy, which means that no crystals at all form. Volcanic glass is not composed of minerals because the magma has cooled too rapidly for crystals to grow, although over time (millions of years) the volcanic glass may crystallize into various silicate minerals.

Mineral Defects

Defects can be introduced in minerals through various means, such as self-radiation. Radiation-induced defects in minerals have been observed long before the discovery of radioactivity. Berzelius observed thermo-luminescence in gadolinite ($(Y,Ce)_2Fe^{2+}Be_2O_2(SiO_4)_2$) caused by α-decay of uranium and thorium trace elements. Unaware of the underlying processes, Baumhauer etched fission-fragment tracks in apatite as early as the nineteenth century.

Classification of Defects

Point (Zero-dimensional) Defects

Point defects are distortions at a single lattice position, such as vacant atomic sites, which are called vacancies. Due to electrical charge neutrality, an equal number of anion and cation vacancies must be present at any time in ionic crystals (Schotky defects). Frenkel defects are a type of vacancy – interstitial pairs in the lattice, which can form if an atom is moved from its lattice site into an interstitial position. In ionic crystals both anion and cation Frenkel defects exist. Point defects can also be impurity elements substituting regular atoms (anions or cations in ionic crystals). Such defects can be introduced by contamination of minerals with foreign species. If the equilibrium charge of the impurity atom is different from the host crystal, vacancies can compensate the charge difference, e.g., Mg^{2+} impurity in NaCl (rock salt) has the microstructure of $Mg^{2+}v_c^-$ (v_c^- is a cation vacancy having a negative charge). The large variety of point defects that can be present in a mineral are illustrated in Figure for NaCl. Point defects are also color centers in dielectric materials which give rise to color in some minerals.

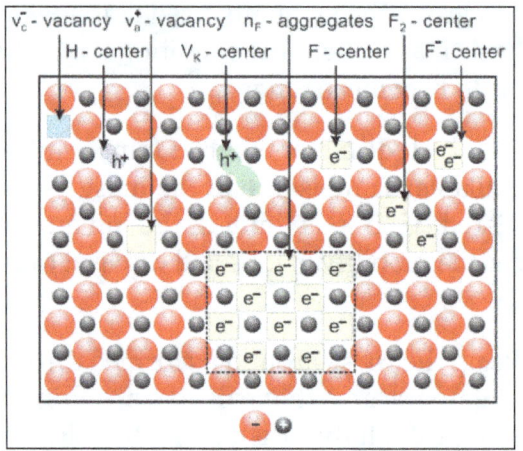

Color centers in NaCl crystals: v_c^- – cation vacancy; H – anion molecule X_2^- replacing an regular anion; v_a^+ – anion vacancy; V_K – anion molecule X_2^- replacing two regular anions (self trsapprd hole); nF – F-center aggregate (quasi Li colloid); F – center ($v_a^+e^-$), F_2 – ($2v_a^+2e^-$); F$^-$ – ($v_a^+2e^-$).

The point-defect concentration depends on temperature. The equilibrium concentration of vacancies (n) is equal to $n/N = \exp(-E_v/k_BT)$, where E_v is the energy to create a vacancy, N is the concentration of lattice atoms (ions), and k_B is the Boltzmann constant. For example, vacancies in NaCl have an E_v of about 2 eV, which corresponds to an anion vacancy concentration of $n_{va} \sim 5 \times 10^{11}cm^{-3}$ at T = 1000 K (close to the melting point).

At elevated temperatures, point defects are mobile and can interact with other defects forming complex defects or defect aggregates. The defect diffusion coefficient is equal to $D(T) = D_0\exp(-Q/k_BT)$, where D_0 is the pre-exponential coefficient and Q the activation energy. Usually the magnitude of Q is larger for vacancies than for interstitials. In NaCl, Q is for both anion and cation vacancies about 2 eV, whereas for neutral chlorine interstitial atoms Q = 0.1 eV.

Line (One-dimensional) Defects

The main line defects within an atomic structure are dislocations. There are two types of dislocations – edge and screw. Edge dislocations are an additional half- plane of atoms introduced into the lattice. Screw dislocations are usually produced by an atomic-step spiral during crystal growth. Mixed edge and screw dislocations are present in many minerals. The main characteristic for dislocations is the Burgers vector (b), which is determined by the so-called Burgers-circuit – the missing link is the vector b. For edge dislocations, b is normal to the dislocation line, and for screw dislocations, b is parallel. The Burgers vector determines the dislocation formation energy (E_{disl}) and is approximately $E_{disl} \sim Gb^2$, where G is the shear modulus (the magnitude for G in various solids is in the range of 10–150 GPa).

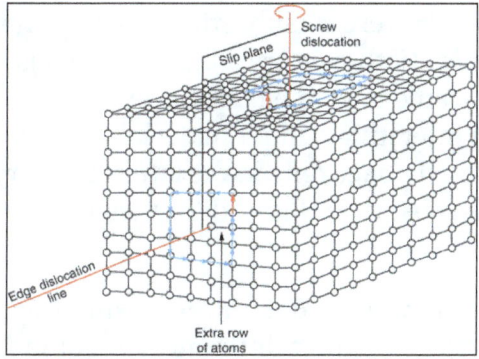

Edge and screw dislocations.

Edge and screw dislocations are created by deviations during growth processes, internal mechanical stresses, and plastic deformations. Dislocations are also mobile at elevated temperatures and can interact with other defects to annihilate or form more complex defect structures. Line defects can be also created by aggregation of point defects (vacancies or interstitials), which occurs, for example, in ionic crystals as a result of particle irradiation.

Planar (Two-dimensional) Defects

The main planar defects are stacking faults, which are defined as local discontinuities in the stacking sequence of the periodic atomic arrangement in the lattice, e.g., ABC in cubic crystals. Details on the structure of stacking faults depend on the crystal symmetry of the host material. Planar defects can be introduced into a material during crystal growth, by interaction of point defects/ dislocations, and by plastic deformation. Stacking faults are more commonly observed in layered minerals.

Bulk (Three-dimensional) Defects

Three dimensional or bulk defects are voids, cracks and impurity clusters. Bulk defects occur on a much larger scale than all other aforementioned crystal defects. Voids are lattice regions with a large number of missing atoms from regular lattice sites, i.e., vacancy aggregates. Voids can form during mineral growth, under irradiation, and plastic deformation.

Finally, impurity atoms of large concentrations can form impurity clusters that lead to local material regions with different phases than the host mineral, the so-called precipitates.

Radiation Damage in Minerals from Self-irradiation

Radiation effects in solids were first observed in colored rock salt (NaCl) and fluorite (CaF_2) minerals. Various minerals can emit a thermo-luminescence signal as a consequence of the presence of radiation-induced defects. Radiation damage in minerals is created through the decay of unstable, natural-occurring radioactive isotopes (^{226}Ra, ^{238}U, ^{232}Th, etc.) that are incorporated in minerals during or after formation. These heavy isotopes were produced by supernova explosions long before the birth of our solar system. The overall level of radiation from these radioisotopes on Earth is currently low. Nevertheless, over geological time intervals, significant amount of radiation damage can accumulate that can lead in extreme cases to the loss of crystallinity of an entire mineral sample.

Radiation-damage studies in solids were mainly stimulated by the need of materials for nuclear applications in the 50s of the twentieth century. There are two general mechanisms for damage creation in solids: atomic displacements and electronic excitations. Atomic displacements lead to the formation of vacancies and interstitials in all materials from metals to insulators, while electronic excitation creates radiation damage only in the latter. Primary atomic displacements induce secondary collisions, and at high absorbed doses, defect aggregates, dislocations and bulk defects are created.

In the 60s of the last century, a new electronic mechanism of Frenkel-defect creation was proposed for ionic crystals. It was observed that different types of radiation (X- or γ-rays, electrons, α-particles, etc.) produce electron–hole pairs and excitons in insulators, which eventually form Frenkel pairs. This electronic or exciton mechanism is active in some dielectric materials if the energy of the electronic excitations, E_{exc}, is higher than the energy required to create a Frenkel pair E_{FD}. The exciton mechanism can explain the formation of structural defects in some materials from initial electronic excitations. Once the Frenkel pairs are created, they can move and produce color centers and, at higher radiation dose, defect aggregates and dislocations. In insulators, the exciton mechanism is much more efficient in producing defects than atomic displacements. Alkali halides and alkaline earth halides have an active exciton mechanism, whereas it is absent in many oxides (MgO, Al_2O_3, Y_2O_3 etc.). Radiation damage can only be created in these oxides through atomic collisions.

Of special interest is damage in minerals that is induced by nuclear fission of uranium ($^{238}U^{92}$) which is as trace element present in many minerals. The spontaneous fission is a radioactive decay event, in which a uranium nucleus splits in two daughter products $^{238}U^{92} \rightarrow {}^{140}Xe^{54} + {}^{96}Sr^{38} + 2^1n^0$ (one example of many possible daughter-product pairs) that carry a high kinetic energy (about 200 MeV in total). The fast fission products slow-down in the mineral lattice and produce a trail of damage, the so-called ion tracks. An ion track is in general a small, cylindrical damage region around the ion path with the length between 10 and 100 μm and a diameter of only a few nanometers.

Radiometric Dating

Radiometric dating is a technique used for mineral or rock samples containing radioactive isotopes. The method is based on a chain of radioactive decays which lead to a stable daughter element, e.g., $^{238}U^{92} \rightarrow {}^{206}Pb^{82}$ or $^{235}U^{92} \rightarrow {}^{207}Pb^{82}$. The radioactive decay has a different rate for various elements and the main parameter for the decay is the half-life time $\tau_{1/2}$. During this time, half of the initial radioactive parent atoms decay. The decay process can be described by the following age equation: $t = \lambda^{-1}\ln(1 + N_f/N_0)$, where t is the age of the mineral, N_f is the number of final isotopes (daughters), N_0 is the number of initial radioactive atoms (parents), and $\lambda = \ln2/\tau_{1/2}$ is the decay constant of the radioactive isotope. The decay constant λ is known for all radioactive isotopes, while the concentration of initial and final isotopes N_0 and N_f must be determined by modern analytical techniques with a high accuracy. It should be noted that the time t only represents the age of the host mineral if the radionuclides were incorporated during mineral formation. Otherwise, this parameter refers to the elapsed time since radionuclide incorporation. If the mineral is exposed to elevated temperatures during some time of the decay of radionuclides, they become mobile and diffuse through the mineral. Above a critical temperature, which depends on the chemical composition of the mineral, the unstable radioactive elements or the stable daughter products will diffuse out of the host mineral, which puts the limit to this type of dating. It was shown that the presence of defects will affect the diffusion process.

For the decay process $^{238}U^{92} \rightarrow ^{206}Pb^{82}$ the half-life is $\tau_{1/2} \approx 4.5 \times 10^9$ years and for $^{235}U^{92} \rightarrow ^{207}Pb^{82}$ $\tau_{1/2} \approx 700 \times 10^6$ years. Thus, these uranium isotopes can be used to determine the age of extremely old minerals and rocks. The oldest mineral dated is zircon, $ZrSiO_4$, having an age of $(4.404 \pm 0.008) \times 10^9$ years. This has led to the conclusion that the oldest minerals formed shortly after the birth of Earth.

To determine ages of much younger geological samples, other isotopes are needed with a significantly shorter half-life, such as those used in the Rb-Sr method with $\tau_{1/2} \approx 50 \times 10^6$ years or ^{14}C with $\tau_{1/2} \approx 5700$ years.

Fission Tracks in Minerals

The principle of fission-track dating is similar to that of other radioactive dating techniques. Unstable $^{238}U^{92}$ nuclei ($\tau_{1/2} \approx 8.4 \times 10^{15}$ years) undergo spontaneous fission and produce two medium-sized daughter products of high kinetic energy (e.g., 80-MeV ^{132}Xe). These fission fragments induce tracks in the host mineral that exhibit a high sensitivity for chemical etching. Chemical etching is used to enlarge fission tracks to micrometer scale (in the form of hollow features) until they are detectable by optical microscopy. This yields the concentration of fission tracks, which is equal to the concentration of daughter products. The estimation of the present concentration of $^{238}U^{92}$ is more complicated and requires modern analytical methods, such as neutron-activation analysis. For this technique, the mineral is bombarded with thermal neutrons that induce nuclear fission of $^{235}U^{92}$, which is easily measured via the detection of fission products. The ratio of $^{238}U^{92}/^{235}U^{92}$ is known and the age of the geological or archeological sample can be determined using the previously described age equation. Fission-track dating is used to date some of the oldest minerals. Since defects anneal at high temperature, tracks recover and shrink in size. Thus, the track-size distribution also contains information about the thermal history of a mineral. If the temperature is high enough to anneal all fission tracks, the clock is reset to "zero." The critical temperatures for zircon, $ZrSiO_4$, and titanate, $CaTiSiO_5$, are about 240 °C and 300 °C, respectively. Fission track formation and annealing under conditions relevant to Earth's interior has been simulated by combining simultaneously irradiation, high temperature and high pressure and applying state-of-the-art materials characterization techniques.

There exists a wide range of defects in minerals from individual point defects to large, three-dimensional defect structures. Defects are induced in minerals through different processes, such as during crystal growth and self-irradiation. Defects diffuse at elevated temperatures and oftentimes form more complex defects or anneal with each other. Defects can dramatically change the physio-chemical properties of a material. One of the most prominent examples is the color of minerals, which can be induced by the underlying defect structure. Finally, defects are of great importance in geochronology, and fission-tracks are, for example, used to obtain the thermal history of archeological and geological samples.

Structural Defects and Color in Minerals

Structure Defects

X-ray diffraction studies, TEM work and HRTEM techniques provide direct evidence that natural minerals contain imperfections at many scales. These structural errors may be at the unit cell scale or macroscopically visible. Such structural imperfections affect basic properties of crystalline materials such as strength, conductivity, mechanical deformation, and color.

Qualitatively, point and line defects and a mosaic of domains separated by defective boundaries. In 3-dimensions, such boundaries would almost certainly be more accurately represented as plane defects.

Point Defects

- Schottky defect- Cation or anion absent from its site in a structure. This needs to be compensated for by either other defects of the opposite charge or the addition of charged species (like an electron).

- Frenkel defect- Absence of an ion from its proper site but its location nearby in an interstitial site. More common for a cation because of its usual smaller size.

- Impurity defect- Addition of an extra ion into the structure.

Line Defects

- Concentrations of defects along linear features - Dislocation.

- Edge dislocation- A plane of atoms that terminates along a line. This kind of error provides a point of weakness for deformation and the defect can migrate through the structure as a slip plane.

- Screw dislocation- Errors along a screw axis that normally is not present in the structure but may be a manifestation of pseudo symmetry or near symmetry inherently present in the space group that the mineral has crystallized within. Such spiral steps are important sites of crystal growth because they provide a good site for the addition of atoms.

Plane Defects

- 2-D zones along which slightly misoriented blocks are joined. The individual blocks may have near perfect short-range order but the crystal as a whole does not have perfect long-range order.

- Stacking fault- HCP interrupted by CCP for example.

Color

Our perception of color in minerals depends on the type of illumination, the mineral itself, and the human eye. Incandescent, fluorescent and sun light are the most common sources and all have different spectral signatures and thus minerals may look different.

Light: Electromagnetic radiation that we can see; very small portion of the total range of possible wavelengths (energy). The shorter the wavelength, the higher the energy. Visible runs from about 375 to 740 nm. We perceive different wavelengths as different colors.

The human eye: Light sensing portion composed of rods and cones. At low light levels we see in shades of gray as detected by the rods which only contain one pigment. At higher levels of illumination, the cones kick in. Each cone contains one of 3 fundamental pigments with maximum absorption in the red, blue or green. Our brain integrates these signals and arrives at an average color. Our eye is not equally sensitive to each of these colors- it is most sensitive to green wavelengths. (This coincides with the peak solar radiation and is therefore an evolutionary development.)

Light and matter: Reflection, refraction, scattering, diffraction, absorbed or transmitted. Part of the energy of the absorbed light can then be emitted as fluorescence.

Main Categories of Interactions Leading to Color

- Dispersed metal ions: Perhaps easiest to understand: single atom.

- Charge-transfer phenomena: Small groups of atoms.

- Color centers: Small groups of atoms.

- Band theory: Large clusters of atoms.

- Physical optics (scattering, diffraction): Large structures.

Dispersed Metal Ions

- An ion absorbs light of a particular energy when that energy matches the amount needed to kick an outer shell electron to a higher energy level. This wavelength is thus preferentially removed from the spectrum and the residual is what we see. The electron normally returns immediately to its ground state but usually does so by giving off some of its energy as heat and the rest as electromagnetic radiation which may or may not lie within our ability to see. If red light was absorbed (blue mineral), then any energy loss will shift the emitted light into the infrared and it will be invisible. In a red mineral (blue absorption), red fluorescence is common and this processes enhances the perceived red color of the mineral.

- The identity of the ion, the valence state, the nature of the neighboring ions and the coordination of the site all effect the absorption capacity of a dispersed metal ion in a mineral. Pleochroism in minerals is caused by different absorption characteristics in different crystallographic directions.

  Example: Rubies (red corundum) and emeralds (green beryl) both owe their colors to Cr+3 in octahedral coordination. The color difference is in the exact position of the absorption band and this hinges on the details of the coordination. (shorter mean Cr-O bond in corundum)

Charge-Transfer

- When an electron jumps from one atom to another:

  ○ Oxygen to metal ion.

  ○ Cation-cation intervalence charge transfer (Fe+2 - Fe+3).

- Aquamarine

- Deep blue of sapphire

Color Centers

- Commonly a result of irradiation by natural or synthetic means:

  ○ Radiation can change the oxidation state of metal ions.

  ○ Interact with defects in the crystal i.e. missing atoms or additional interstitial atoms.

  ○ The removed electron can find a home in one of the defects.

- Smoky quartz is formed by removing an electron from an Al ion that had substituted for a Si. Heating allows the electrons to come home and the smoky color will disappear.

- Amethyst is formed when Fe+3 substituting for Si is ionized to Fe+4 by radiation. The deep purple color is due to O-2 -> Fe+4 charge transfer which is centered in the yellow-green portion of the spectrum.

- Sodalite (hackmanite): Electron in Cl- hole in a Na tetrahedron.

Band Theory

Electrons that can be delocalized over the entire crystal; into an electronic energy band composed of many many closely spaced energy levels. There are 2 bands in such materials, a low energy valence band that is fully populated and a high energy conduction band that is normally empty. 3 end-member scenarios exist:

- If the band gap is greater than the maximum energy of visible light then no transitions occur, no visible absorbance occurs and the mineral is transparent. Such minerals are inherently electrical insulators.

- If the gap is less than the energy of violet light, the high energy end of the spectrum tends to be absorbed leaving the red- this is the cause of the red color of cinnabar.

- If the gap is less than all the energies represented by visible light then the whole spectrum is absorbed. The mineral commonly appears black and opaque. All metals have this property (or no gap at all). Metals appear shiny (metallic luster) because the electrons quickly return to their original energy level, emitting the same energy they absorbed. In cases where some wavelengths are absorbed emitted more efficiently than others, a color is produced (gold vs platinum).

- Nitrogen in diamonds -> yellow.

- Boron in diamonds -> blue.

Physical Phenomena

- Interference caused by thin films - pearls.

- Diffraction: caused by regular 'layers' on the scale of the wavelength of light - opal:

  - 250nm spheres diffract red light.

  - Spheres down to 140nm diffract the other colors.

- Scattering: From particles smaller than the wavelength of light: reason the sky is blue during the day and red at sunrise and sunset (blue is scattered more efficiently than red).

- Opalescence caused by spheres too small to diffract.

Mineral Stability and Phase Diagrams

There are four major processes by which minerals form. Each of these occurs within a limited range of environmental conditions. First, the chemical ingredients must be present, and second, the pressure and temperature conditions must be right.

- Precipitation from a fluid like H_2O or CO_2:

  - Hydrothermal Processes - T = 100 - 500°C, P = 0 to 1000 MPa (10 kb).

  - Diagenesis - T = 0 - 200°C, P = 1 atm - 300 MPa (3kb).

  - Evaporation - T = 10 - 40°C, P = 1atm

  - Weathering - T = 10 - 100°C, P = 1 atm - 10 MPa (0.1 kb)

  - Biological activity - T = 10 - 40 °C, P = 1atm - 1Mpa.(0.01kb)

- Sublimation from a vapor. This process is somewhat more rare, but can take place at a volcanic vent, or deep in space where the pressure is near vacuum. T = 0 - 500 °C, P = 0 - 1 atm.

- Crystallization from a liquid. This takes place during crystallization of molten rock (magma) either below or at the Earth's surface. Results in igneous rocks, T = 600 - 1300 °C, P = 1atm - 3,000 MPa (30kb).

- Solid - Solid reactions. This process involves minerals reacting with other minerals in the solid state to produce one or more new minerals:

  ○ Diagenesis - T = 100 - 200 °C, P = 1 atm - 300 MPa.

  ○ Metamorphism - T = 200 °C - melting T, P = 300 - 1000 MPa.

Thus, for any given system we can define temperature, pressure, and compositional variables that determine what minerals are stable. An understanding of mineral stability is essential in understanding which minerals form, and allow us to determine the conditions present when we encounter minerals in the Earth.

Both temperature and pressure vary with depth in the Earth. Pressure is related to depth because pressure is caused by the weight of the overlying rocks. The way that pressure and temperature vary in the Earth is called the Geothermal Gradient.

The average, or sometimes called normal, geothermal gradient in the upper part of the Earth is about 25 °C/km. But, the geothermal gradient can vary from 200 °C/km in areas where hot igneous bodies are intruding at shallow levels of the crust to 10 °C/km, in areas like subduction zones where cold lithosphere descends back into the mantle.

Geothermal gradients deeper in the earth become much lower than those near the surface.

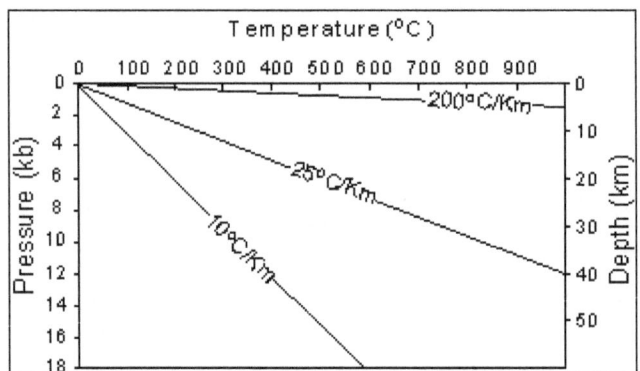

Phase Diagrams

A phase diagram is a graphical representation of chemical equilibrium. Since chemical equilibrium is dependent on the composition of the system, the pressure, and the temperature, a phase diagram should be able to tell us what phases are in equilibrium for any composition at any temperature and pressure of the system. First, a few terms will be defined.

System - A system is that part of the universe which is under consideration. Thus, it may or may not have fixed boundaries, depending on the system. For example, if we are experimenting with a beaker containing salt and water, and all we are interested in is the salt and water contained in that beaker, then our system consists only of salt and water contained in the beaker.

If the system cannot exchange mass or energy with its surroundings, then it is termed an isolated system. (Our salt and water system, if we put a lid on it to prevent evaporation, and enclosed it in a perfect thermal insulator to prevent it from heating or cooling, would be an isolated system.)

If the system can exchange energy, but not mass with its surroundings, we call it a closed system. (Our beaker, still sealed, but without the thermal insulator is a closed system).

If the system can exchange both mass and energy with its surroundings, we call it an open system. (Our beaker - salt - water system open to the air and not insulated is thus an open system).

Phase - A phase is a physically separable part of the system with distinct physical and chemical properties. A system must consist of one or more phases. For example, in our salt-water system, if all of the salt is dissolved in the water, consists of only one phase (a sodium chloride - water solution). If we have too much salt, so that it cannot all dissolve in the water, we have 2 phases, the sodium chloride - water solution and the salt crystals. If we heat our system under sealed conditions, we might have 3 phases, a gas phase consisting mostly of water vapor, the salt crystals, and the sodium chloride - water solution.

In a magma a few kilometers deep in the earth we might expect one or more phases. For example if it is very hot so that no crystals are present, and there is no free vapor phase, the magma consists of one phase, the liquid. At lower temperature it might contain a vapor phase, a liquid phase, and one or more solid phases. For example, if it contains crystals of plagioclase and olivine, these two minerals would be considered as two separate solid phases because olivine is physically and chemically distinct from plagioclase.

Component - Each phase in the system may be considered to be composed of one or more components. The number of components in the system must be the minimum required to define all of the phases. For example, in our system salt and water, we might have the components Na, Cl, H, and O (four components), NaCl, H, and O (three components), NaCl and HO (two components), or Na-Cl-H_2O (one component). However, the possible phases in the system can only consist of crystals of halite (NaCl), H_2O either liquid or vapor, and NaCl-H_2O solution. Thus only two components (NaCl and H_2O) are required to define the system, because the third phase (NaCl - H_2O solution) can be obtained by mixing the other two components.

Phase Rule

The phase rule is an expression of the number of variables and equations that can be used to describe a system in equilibrium. In simple terms, the number of variables are the number of chemical components in the system plus the extensive variables, temperature and pressure. The number of phases present will depend on the variance or degrees of freedom of the system.

The general form of the phase rule is stated as follows.

$$F = C + 2 - P$$

Here F is the number of degrees of freedom or variance of the system.

C is the number of components, as defined above, in the system.

P is the number of phases in equilibrium, and the 2 comes from the two extensive variables, Pressure and Temperature.

To see how the phase rule works, let's start with a simple one component system - the system Al_2SiO_5, shown in the Pressure, Temperature phase diagram below.

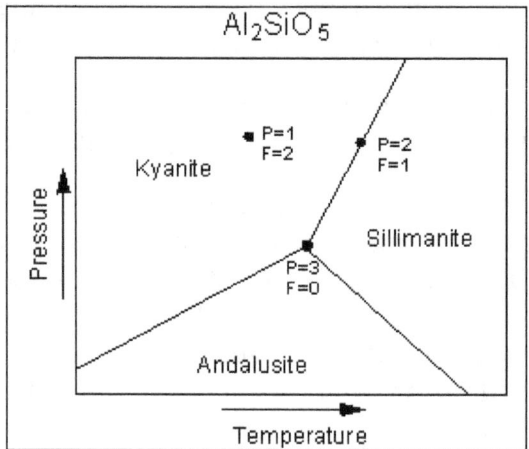

First look at the point in the field of kyanite stability. Since kyanite is the only phase present, P=1. F is 2 at this point, because one could change both temperature and pressure by small amounts without affecting the number of phases present. We say that this area of kyanite stability on the phase diagram is a divariant field (variance, F =2).

Next look at the point on the phase boundary between kyanite and sillimanite. For any point on such a boundary the number of phases, P, will be 2. Using the phase rule we find that F = 1, or there is one degree of freedom. This means there is only one independent variable.

If we change pressure, temperature must also change in order to keep both phases stable. The phase assemblage is said to be univariant in this case, and the phase boundaries are univariant lines (or curves in the more general case).

Finally, we look at the point where all three univariant lines intersect. At this point, 3 phases, kyanite, andalusite, and sillimanite all coexist at equilibrium. Note that this is the only point where all three phases can coexist. For this case, P=3, and F, from the phase rule, is 0. There are no degrees of freedom, meaning that any change in pressure or temperature will result in a change in the number of phases. The three phase assemblage in a one component system is said to be invariant.

Formation of Minerals

In order for a mineral crystal to grow, the elements needed to make it must be present in the appropriate proportions, the physical and chemical conditions must be favourable, and there must be sufficient time for the atoms to become arranged.

Physical and chemical conditions include factors such as temperature, pressure, presence of water, pH, and amount of oxygen available. Time is one of the most important factors because it takes time for atoms to become ordered. If time is limited, the mineral grains will remain very small. The presence of water enhances the mobility of ions and can lead to the formation of larger crystals over shorter time periods.

Most of the minerals that make up the rocks around us formed through the cooling of molten rock, known as magma. At the high temperatures that exist deep within Earth, some geological materials are liquid. As magma rises up through the crust, either by volcanic eruption or by more gradual processes, it cools and minerals crystallize. If the cooling process is rapid (minutes, hours, days, or years), the components of the minerals will not have time to become ordered and only small crystals can form before the rock becomes solid. The resulting rock will be fine-grained (i.e., crystals less than 1 mm). If the cooling is slow (from decades to millions of years), the degree of ordering will be higher and relatively large crystals will form. In some cases, the cooling will be so fast (seconds) that the texture will be glassy, which means that no crystals at all form. Volcanic glass is not composed of minerals because the magma has cooled too rapidly for crystals to grow, although over time (millions of years) the volcanic glass may crystallize into various silicate minerals.

Minerals can also form in several other ways:

- Precipitation from aqueous solution (i.e., from hot water flowing underground, from evaporation of a lake or inland sea, or in some cases, directly from seawater).

- Precipitation from gaseous emanations.

- Metamorphism — Formation of new minerals directly from the elements within existing minerals under conditions of elevated temperature and pressure.

- Weathering — During which minerals unstable at Earth's surface may be altered to other minerals.

- Organic formation — Formation of minerals within shells (primarily calcite) and teeth and bones (primarily apatite) by organisms (these organically formed minerals are still called minerals because they can also form inorganically).

Opal is a mineraloid, because although it has all of the other properties of a mineral, it does not have a specific structure. Pearl is not a mineral because it can only be produced by organic processes.

Mineral Nucleation

Nucleation is the first step in the formation of either a new thermodynamic phase or a new structure via self-assembly or self-organization. Nucleation is typically defined to be the process that determines how long an observer has to wait before the new phase or self-organized structure appears. For example, if a volume of water is cooled (at atmospheric pressure) below 0° C, it will tend to freeze into ice. Volumes of water cooled only a few degrees below 0° C often stay completely free of ice for long periods. At these conditions, nucleation of ice is either slow or does not occur at all. However, at lower temperatures ice crystals appear after little or no delay. At these conditions ice nucleation is fast. Nucleation is commonly how first-order phase transitions start, and then it is the start of the process of forming a new thermodynamic phase. By contrast, new phases at continuous phase transitions start to form immediately.

Nucleation is often found to be very sensitive to impurities in the system. These impurities may be too small to be seen by the naked eye, but still can control the rate of nucleation. Because of this, it is often important to distinguish between heterogeneous nucleation and homogeneous nucleation. Heterogeneous nucleation occurs at *nucleation sites* on surfaces in the system. Homogeneous nucleation occurs away from a surface.

Characteristics

Nucleation is usually a stochastic (random) process, so even in two identical systems nucleation will occur at different times. This behaviour is similar to radioactive decay. A common mechanism is illustrated in the animation to the right. This shows nucleation of a new phase (shown in red) in an existing phase (white). In the existing phase microscopic fluctuations of the red phase appear and decay continuously, until an unusually large fluctuation of the new red phase is so large it is more favourable for it to grow than to shrink back to nothing. This nucleus of the red phase then grows and converts the system to this phase. The standard theory that describes this behaviour for the nucleation of a new thermodynamic phase is called classical nucleation theory. However, the CNT fails in describing experimental results of vapour to liquid nucleation even for model substances like Argon by several orders of magnitude.

For nucleation of a new thermodynamic phase, such as the formation of ice in water below 0 °C, if the system is not evolving with time and nucleation occurs in one step, then the probability that nucleation has *not* occurred should undergo exponential decay as seen in radioactive decay. This is seen for example in the nucleation of ice in supercooled small water droplets. The decay rate of the exponential gives the nucleation rate. Classical nucleation theory is a widely used approximate theory for estimating these rates, and how they vary with variables such as temperature. It correctly predicts that the time you have to wait for nucleation decreases extremely rapidly when supersaturated.

It is not just new phases such as liquids and crystals that form via nucleation followed by growth. The self-assembly process that forms objects like the amyloid aggregates associated with Alzheimer's disease also starts with nucleation. Energy consuming self-organising systems such as the microtubules in cells also show nucleation and growth.

Heterogeneous Nucleation often Dominates Homogeneous Nucleation

Heterogeneous nucleation, nucleation with the nucleus at a surface, is much more common than homogeneous nucleation. For example, in the nucleation of ice from supercooled water droplets, purifying the water to remove all or almost all impurities results in water droplets that freeze below around - 35 °C, whereas water that contains impurities may freeze at - 5 °C or warmer. Thus here, we have direct evidence that nucleation of ice on impurities can occur at much higher temperatures (smaller degrees of supercooling) than without impurities.

This observation that heterogeneous nucleation can occur when the rate of homogeneous nucleation is essentially zero, is often understood using classical nucleation theory. This predicts that the nucleation slows exponentially with the height of a free energy barrier ΔG^*. This barrier comes from the free energy penalty of forming the surface of the growing nucleus. For homogeneous nucleation the nucleus is approximated by a sphere, but as we can see in the schematic of macroscopic droplets to the right, droplets on surfaces are not complete spheres and so the area of the interface between the droplet and the surrounding fluid is less than a sphere's $4\pi r^2$. This reduction in surface area of the nucleus reduces the height of the barrier to nucleation and so speeds nucleation up exponentially.

Nucleation can also start at the surface of a liquid. For example, computer simulations of gold nanoparticles show that the crystal phase nucleates at the liquid-gold surface.

Computer Simulation Studies of Simple Models

Classical nucleation theory makes a number of assumptions, for example it treats a microscopic nucleus as if it is a macroscopic droplet with a well-defined surface whose free energy is estimated using an equilibrium property: the interfacial tension σ. For a nucleus that may be only of order ten molecules across it is not always clear that we can treat something so small as a volume plus a surface. Also nucleation is an inherently out of thermodynamic equilibrium phenomenon so it is not always obvious that its rate can be estimated using equilibrium properties.

However, modern computers are powerful enough to calculate essentially exact nucleation rates for simple models. These have been compared with the classical theory, for example for the case of nucleation of the crystal phase in the model of hard spheres. This is a model of perfectly hard spheres in thermal motion, and is a simple model of some colloids. For the crystallization of hard spheres the classical theory is a very reasonable approximate theory. So for the simple models we can study, classical nucleation theory works quite well, but we do not know if it works equally well for (say) complex molecules crystallising out of solution.

Spinodal Region

Phase-transition processes can also be explained in terms of spinodal decomposition, where phase separation is delayed until the system enters the unstable region where a small perturbation in composition leads to a decrease in energy and, thus, spontaneous growth of the perturbation. This region of a phase diagram is known as the spinodal region and the phase separation process is known as spinodal decomposition and may be governed by the Cahn–Hilliard equation.

Nucleation of Crystals

In many cases, liquids and solutions can be cooled down or concentrated up to conditions where the liquid or solution is significantly less thermodynamically stable than the crystal, but where no crystals will form for minutes, hours, weeks or longer. Nucleation of the crystal is then being prevented by a substantial barrier. This has consequences, for example cold high altitude clouds may contain large numbers of small liquid water droplets that are far below 0 °C.

In small volumes, such as in small droplets, only one nucleation event may be needed for crystallisation. In these small volumes, the time until the first crystal appears is usually defined to be the nucleation time. In larger volumes many nucleation events will occur. A simple model for crystallisation in that case, that combines nucleation and growth is the KJMA or Avrami model.

Primary and Secondary Nucleation

The time until the appearance of the first crystal is also called primary nucleation time, to distinguish it from secondary nucleation times. Primary here refers to the first nucleus to form, while secondary nuclei are crystal nuclei produced from a preexisting crystal. Primary nucleation describes the transition to a new phase that does not rely on the new phase already being present, either because it is the very first nucleus of that phase to form, or because the nucleus forms far from any pre-existing piece of the new phase. Particularly in the study of crystallisation, secondary nucleation can be important. This is the formation of nuclei of a new crystal directly caused by pre-existing crystals.

For example, if the crystals are in a solution and the system is subject to shearing forces, small crystal nuclei could be sheared off a growing crystal, thus increasing the number of crystals in the system. So both primary and secondary nucleation increase the number of crystals in the system but their mechanisms are very different, and secondary nucleation relies on crystals already being present.

Experimental Observations on the Nucleation Times for the Crystallisation of Small Volumes

It is typically difficult to experimentally study the nucleation of crystals. The nucleus is microscopic, and thus too small to be directly observed. In large liquid volumes there are typically multiple nucleation events, and it is difficult to disentangle the effects of nucleation from those of growth of the nucleated phase. These problems can be overcome by working with small droplets. As nucleation is stochastic, many droplets are needed so that statistics for the nucleation events can be obtained.

Nucleation occurs in different droplets at different times, hence the fraction is not a simple step function that drops sharply from one to zero at one particular time. The red curve is a fit of a Gompertz function to the data. This is a simplified version of the model Pound and La Mer used to model their data. The model assumes that nucleation occurs due to impurity particles in the liquid tin droplets, and it makes the simplifying assumption that all impurity particles produce nucleation at the same rate. It also assumes that these particles are Poisson distributed among the liquid tin droplets. The fit values are that the nucleation rate due to a single impurity particle is 0.02/s, and the average number of impurity particles per droplet is 1.2. Note that about 30% of the tin droplets never freeze; the data plateau at a fraction of about 0.3. Within the model this is assumed to be because, by chance, these droplets do not have even one impurity particle and so there is no heterogeneous nucleation. Homogeneous nucleation is assumed to be negligible on the timescale of this experiment. The remaining droplets freeze in a stochastic way, at rates 0.02/s if they have one impurity particle, 0.04/s if they have two, and so on.

These data are just one example, but they illustrate common features of the nucleation of crystals in that there is clear evidence for heterogeneous nucleation, and that nucleation is clearly stochastic.

Ice

The freezing of small water droplets to ice is an important process, particularly in the formation and dynamics of clouds. Water (at atmospheric pressure) does not freeze at 0 °C, but rather at temperatures that tend to decrease as the volume of the water decreases and as the water impurity increases.

Thus small droplets of water, as found in clouds, may remain liquid far below 0 °C.

An example of experimental data on the freezing of small water droplets is shown at the right. The plot shows the fraction of a large set of water droplets, that are still liquid water, i.e., have not yet frozen, as a function of temperature. Note that the highest temperature at which any of the droplets freezes is close to -19 °C, while the last droplet to freeze does so at almost -35 °C.

Examples of the Nucleation of Fluids (Gases and Liquids)

- Clouds form when wet air cools (often because the air rises) and many small water droplets nucleate from the supersaturated air. The amount of water vapor that air can carry decreases with lower temperatures. The excess vapor begins to nucleate and to form small water droplets which form a cloud. Nucleation of the droplets of liquid water is heterogeneous, occurring on particles referred to as cloud condensation nuclei. Cloud seeding is the process of adding artificial condensation nuclei to quicken the formation of clouds.

- Bubbles of carbon dioxide *nucleate* shortly after the pressure is released from a container of carbonated liquid.

- Nucleation in boiling can occur in the bulk liquid if the pressure is reduced so that the liquid becomes superheated with respect to the pressure-dependent boiling point. More often, nucleation occurs on the heating surface, at *nucleation sites*. Typically, nucleation sites are tiny crevices where free gas-liquid surface is maintained or spots on the heating surface with lower wetting properties. Substantial superheating of a liquid can be achieved after the liquid is de-gassed and if the heating surfaces are clean, smooth and made of materials well wetted by the liquid.

- Some champagne stirrers operate by providing many nucleation sites via high surface-area and sharp corners, speeding the release of bubbles and removing carbonation from the wine.

- The Diet Coke and Mentos eruption offers another example. The surface of Mentos candy provides nucleation sites for the formation of carbon-dioxide bubbles from carbonated soda.

- Both the bubble chamber and the cloud chamber rely on nucleation, of bubbles and droplets, respectively.

Examples of the Nucleation of crystals

- The most common crystallisation process on Earth is the formation of ice. Liquid water does not freeze at 0 °C unless there is ice already present, cooling significantly below 0 °C is required to nucleate ice and so for the water to freeze. For example, small droplets of very pure water can remain liquid down to below -30 °C although ice is the stable state below 0 °C.

- Many of the materials we make and use are crystalline, but are made from liquids, eg crystalline iron made from liquid iron cast into a mold. So the nucleation of crystalline materials is widely studied in industry. It is used heavily in the chemical industry for cases such as in the preparation of metallic ultradispersed powders that can serve as catalysts. For example, platinum deposited onto TiO_2 nanoparticles catalyses the liberation of hydrogen from water. It is an important factor in the semiconductor industry, as the band gap energy in semiconductors is influenced by the size of nanoclusters.

Crystal Twinning

Crystal twinning occurs when two separate crystals share some of the same crystal lattice points in a symmetrical manner. The result is an intergrowth of two separate crystals in a variety of specific configurations. The surface along which the lattice points are shared in twinned crystals is called a composition surface or twin plane.

Crystallographers classify twinned crystals by a number of twin laws. These twin laws are specific to the crystal system. The type of twinning can be a diagnostic tool in mineral identification. Twinning is an important mechanism for permanent shape changes in a crystal.

Twinning can often be a problem in X-ray crystallography, as a twinned crystal does not produce a simple diffraction pattern.

Twin Laws

Twin laws are either defined by their twin planes (i.e. {hkl}) or the direction of the twin axes (i.e. [hkl]). If the twin law can be defined by a simple planar composition surface, the twin plane is always parallel to a possible crystal face and never parallel to an existing plane of symmetry (remember that twinning adds symmetry).

If the twin law is a rotation axis, the composition surface will be irregular, the twin axis will be perpendicular to a lattice plane, but will never be an even-fold rotation axis of the existing symmetry. For example, twinning cannot occur on a new 2 fold axis that is parallel to an existing 4-fold axis.

Common Twin Laws

In the isometric system, the most common types of twins are the Spinel Law (twin plane, parallel to an octahedron), where the twin axis is perpendicular to an octahedral face, and the Iron Cross which is the interpenetration of two pyritohedrons a subtype of dodecahedron.

In the hexagonal system, calcite shows the contact twin laws {0001} and {0112}. Quartz shows the Brazil Law {1120}, and Dauphiné Law which are penetration twins caused by transformation and Japanese Law {1122} which is often caused by accidents during growth.

In the tetragonal system, cyclical contact twins are the most commonly observed type of twin, such as in rutile titanium dioxide and cassiterite tin oxide.

In the orthorhombic system, crystals usually twin on planes parallel to the prism face, where the most common is a {110} twin which produces cyclical twins, such as in aragonite, chrysoberyl, and cerussite.

In the monoclinic system, twin occur most often on the planes {100} and {001} by the Manebach Law {001}, Carlsbad Law [001], Braveno Law {021} in orthoclase, and the Swallow Tail Twins {001} in gypsum.

In the triclinic system, the most commonly twinned crystals are the feldspar minerals plagioclase and microcline.

Types of Twinning

Simple twinned crystals may be contact twins or penetration twins. *Contact twins* share a single composition surface often appearing as mirror images across the boundary. Plagioclase, quartz, gypsum, and spinel often exhibit contact twinning. *Merohedral twinning* occurs when the lattices of the contact twins superimpose in three dimensions, such as by relative rotation of one twin from the other. An example is metazeunerite. In *penetration twins* the individual crystals have the appearance of *passing through* each other in a symmetrical manner. Orthoclase, staurolite, pyrite, and fluorite often show penetration twinning.

If several twin crystal parts are aligned by the same twin law they are referred to as *multiple* or *repeated twins*. If these multiple twins are aligned in parallel they are called *polysynthetic twins*. When the multiple twins are not parallel they are *cyclic twins*. Albite, calcite, and pyrite often show polysynthetic twinning. Closely spaced polysynthetic twinning is often observed as striations or fine parallel lines on the crystal face. Rutile, aragonite, cerussite, and chrysoberyl often exhibit cyclic twinning, typically in a radiating pattern. But in general, based on the relationship between the twin axis and twin plane, there are 3 types of twinning:

- Parallel twinning, when the twin axis and compositional plane lie parallel to each other,

- Normal twining, when the twin plane and compositional plane lie normally, and

- Complex twining, a combination of parallel twinning and normal twinning on one compositional plane.

Modes of Formation

There are three modes of formation of twinned crystals. Growth twins are the result of an interruption or change in the lattice during formation or growth due to a possible deformation from a larger substituting ion. Annealing or transformation twins are the result of a change in crystal system during cooling as one form becomes unstable and the crystal structure must re-organize or transform into another more stable form. Deformation or gliding twins are the result of stress on the crystal after the crystal has formed. If a metal with face-centered cubic (fcc) structure, like Al, Cu, Ag, Au, etc., is subjected to stress, it will experience twinning. The formation and migration of twin boundaries is partly responsible for ductility and malleability of fcc metals.

Deformation twinning is a common result of regional metamorphism. Crystal twinning is also used as an indicator of force direction in mountain building processes in orogeny research.

Crystals that grow adjacent to each other may be aligned to resemble twinning. This parallel growth simply reduces system energy and is not twinning.

Mechanisms of Formation

Twinning can occur by cooperative displacement of atoms along the face of the twin boundary. This displacement of a large quantity of atoms simultaneously requires significant energy to perform. Therefore, the theoretical stress required to form a twin is quite high. It is believed that twinning is associated with dislocation motion on a coordinated scale, in contrast to slip, which is caused by independent glide at several locations in the crystal.

Twinning and slip are competitive mechanisms for crystal deformation. Each mechanism is dominant in certain crystal systems and under certain conditions. In fcc metals, slip is almost always dominant because the stress required is far less than twinning stress.

Compared to slip, twinning produces a deformation pattern that is more heterogeneous in nature. This deformation produces a local gradient across the material and near intersections between twins and grain boundaries. The deformation gradient can lead to fracture along the boundaries, particularly in bcc transition metals at low temperatures.

Deposition of Twins

The conditions of crystal formation in solution have an effect on the type and density of dislocations in the crystal. It frequently happens that the crystal is oriented so that there will a more rapid deposition of material on one part than on another; for instance, if the crystal be attached to some other solid it cannot grow in that direction. If the crystal is freely suspended in the solution and material for growth is supplied at the same rate on all sides does an equally developed form result.

Twin Boundaries

Twin boundaries occur when two crystals of the same type intergrow so that only a slight misorientation exists between them. It is a highly symmetrical interface, often with one crystal the mirror image of the other; also, atoms are shared by the two crystals at regular intervals. This is also a much lower-energy interface than the grain boundaries that form when crystals of arbitrary orientation grow together. Twin boundaries may also display a higher degree of symmetry than the single crystal. These twins are called *mimetic* or *pseudo-symmetric* twins.

Twin boundaries are partly responsible for shock hardening and for many of the changes that occur in cold work of metals with limited slip systems or at very low temperatures. They also occur due to martensitic transformations: the motion of twin boundaries is responsible for the pseudoelastic and shape-memory behavior of nitinol, and their presence is partly responsible for the hardness due to quenching of steel. In certain types of high strength steels, very fine deformation twins act as primary obstacles against dislocation motion. These steels are referred to as 'TWIP' steels, where TWIP stands for *twinning-induced plasticity*.

Appearance in Different Structures

Of the three common crystalline structures bcc, fcc, and hcp, the hcp structure is the most likely to form deformation twins when strained, because they rarely have a sufficient number of slip systems for an arbitrary shape change. High strain rates, low stacking-fault energy and low temperatures facilitate deformation twinning.

References

- Sear, R.P. (2007). "Nucleation: theory and applications to protein solutions and colloidal suspensions" (PDF). Journal of Physics: Condensed Matter. 19 (3): 033101. Bibcode:2007JPCM...19c3101s. Citeseerx 10.1.1.605.2550. Doi:10.1088/0953-8984/19/3/033101

- Formation-of-minerals, chapter, geology: opentextbc.ca, Retrieved 28 April, 2019

- Kelton, Ken; Greer, Alan Lindsay (2010). Nucleation in Condensed Matter: Applications in Materials and Biology. Amsterdam: Elsevier Science & Technology. ISBN 9780080421476.

- Mineral-stability, sanelson: tulane.edu, Retrieved 29 June, 2019

- Gillam, J.E.; macphee, C.E. (2013). "Modelling amyloid fibril formation kinetics: mechanisms of nucleation and growth". Journal of Physics: Condensed Matter. 25 (37): 373101

- Steinmetz, D.R.; Jäpel, T.; Wietbrock, B.; Eisenlohr, P.; Gutierrez-Urrutia, I.; Saeed (2013), "Revealing the strain-hardening behavior of twinning-induced plasticity steels: Theory, simulations, experiments", Acta Materialia, 61 (2): 494, doi:10.1016/j.actamat.2012.09.064

Crystallography 5

- **Crystal**

- **Lattice Plane**

- **Miller Index**

- **Crystallographic Restriction Theorem**

- **X-ray Crystallography**

- **Crystallization**

The branch of science which studies the arrangement of atoms within a crystalline solid is known as crystallography. Some of the important areas of study within this field are crystal structure, crystal forms and crystallographic effect. This chapter discusses these focus areas of crystallography in detail.

Crystallography is the science that examines crystals, which can be found everywhere in nature—from salt to snowflakes to gemstones. Crystallographers use the properties and inner structures of crystals to determine the arrangement of atoms and generate knowledge that is used by chemists, physicists, biologists, and others. Within the past century, crystallography has been a primary force in driving major advances in the detailed understanding of materials, synthetic chemistry, the understanding of basic principles of biological processes, genetics, and has contributed to major advances in the development of drugs for numerous diseases.

Crystallographers use X-ray, neutron, and electron diffraction techniques to identify and characterize solid materials. They commonly bring in information from other analytical techniques, including X-ray fluorescence, spectroscopic techniques, microscopic imaging, and computer modeling and visualization to construct detailed models of the atomic arrangements in solids. This provides valuable information on a material's chemical makeup, polymorphic form, defects or disorder, and electronic properties. It also sheds light on how solids perform under temperature, pressure, and stress conditions.

Crystal-growing specialists use a variety of techniques to produce crystalline forms of compounds for use in research or manufacturing. They may be experts in working with hard-to-crystallize materials, or they may grow crystals to exacting specifications for use in computer chips, solar cells, optical components, or pharmaceutical products.

Single-crystal X-ray crystallography is widely considered to be the gold standard for establishing the structures of crystalline solids. This method is used to establish patent claims, establish structure-property relationships for new compounds, and many other applications. However, powder crystallography instrumentation and data analysis software have emerged over the past 30 years as powerful methods for investigating the structures of materials that cannot be studied with single-crystal methods. Powder methods are used in a wide variety of investigations, including forensic analyses, identifying components of mixtures, and identifying properties of polymers and other poorly crystallized materials.

The pharmaceutical and biochemical fields rely extensively on crystallographic studies. Proteins and other biological materials (including viruses) may be crystallized to aid in studying their structures and composition. Many important pharmaceuticals are administered in crystalline form, and detailed descriptions of their crystal structures provide evidence to verify claims in patents.

Instrument manufacturers hire crystallographers for customer sales and support functions, including instrument repair and helping customers with special projects. Staff crystallographers at the national laboratories develop and maintain leading-edge research instruments and software capabilities. They also assist visiting users in setting up and running experiments using specialized techniques, including synchrotron X-ray diffraction and neutron diffraction. Universities employ staff members to maintain and operate their research laboratories and to train students to use the instruments.

Some crystallographers develop instrumentation and software for collecting, analyzing, and visualizing data and for translating this data into crystal structure models. Some crystallographers maintain and develop archival databases at industrial and academic institutions, as well as some nonprofits and government laboratories.

Service laboratories hire diffraction technicians to prepare and catalog samples, run the data collections, and prepare routine reports on the results. Technicians may also be called on to perform routine instrument maintenance and simple repairs.

Forensics laboratories use crystallography to investigate cases involving product adulteration or counterfeiting. They may identify minerals, metals, or other materials found at crime scenes. They may also identify corrosion products and other residues found at the site of an industrial accident to help verify the events leading up to the accident.

An image of a small object is made using a lens to focus the beam, similar to a lens in a microscope. However, the wavelength of visible light (about 4000 to 7000 Å) is three orders of magnitude longer than the length of typical atomic bonds and atoms themselves (about 1 to 2 Å). Therefore, obtaining information about the spatial arrangement of atoms requires the use of radiation with shorter wavelengths, such as X-ray or neutron beams. Employing shorter wavelengths implied abandoning microscopy and true imaging, however, because there exists no material from which a lens capable of focusing this type of radiation can be created. Scientists have had some success focusing X-rays with microscopic Fresnel zone plates made from gold, and by critical-angle reflection inside long tapered capillaries. Diffracted X-ray or neutron beams cannot be focused to produce images, so the sample structure must be reconstructed from the diffraction pattern. Sharp features in the diffraction pattern arise from periodic, repeating structure in the sample, which are often very strong due to coherent reflection of many photons from many regularly spaced in-stances of similar structure, while non-periodic components of the structure result in diffuse (and usually weak) diffraction features - areas with a higher density and repetition of atom order tend to reflect more light toward one point in space when compared to those areas with fewer atoms and less repetition.

Because of their highly ordered and repetitive structure, crystals give diffraction patterns of sharp Bragg reflection spots, and are ideal for analyzing the structure of solids.

Notation

- Coordinates in square brackets such as denote a direction vector (in real space).

- Coordinates in angle brackets or chevrons such as <100> denote a family of directions which are related by symmetry operations. In the cubic crystal system for example, <100> would mean [100], [010], or the negative of any of those directions.

- Miller indices in parentheses such as (100) denote a plane of the crystal structure, and regular repetitions of that plane with a particular spacing. In the cubic system, the normal to the (hkl) plane is the direction [hkl], but in lower-symmetry cases, the normal to (hkl) is not parallel to [hkl].

- Indices in curly brackets or braces such as {100} denote a family of planes and their normals. In cubic materials the symmetry makes them equivalent, just as the way angle brackets denote a family of directions. In non-cubic materials, <hkl> is not necessarily perpendicular to {hkl}.

Techniques

Some materials that have been analyzed crystallographically, such as proteins, do not occur naturally as crystals. Typically, such molecules are placed in solution and allowed to slowly crystallize through vapor diffusion. A drop of solution containing the molecule, buffer, and precipitants is sealed in a container with a reservoir containing a hygroscopic solution. Water in the drop diffuses to the reservoir, slowly increasing the concentration and allowing a crystal to form. If the concentration were to rise more quickly, the molecule would simply precipitate out of solution, resulting in disorderly granules rather than an orderly and hence usable crystal.

Once a crystal is obtained, data can be collected using a beam of radiation. Although many universities that engage in crystallographic research have their own X-ray producing equipment, synchrotrons are often used as X-ray sources, because of the purer and more complete patterns such sources can generate. Synchrotron sources also have a much higher intensity of X-ray beams, so data collection takes a fraction of the time normally necessary at weaker sources. Complementary neutron crystallography techniques are used to identify the positions of hydrogen atoms, since X-rays only interact very weakly with light elements such as hydrogen.

Producing an image from a diffraction pattern requires sophisticated mathematics and often an iterative process of modelling and refinement. In this process, the mathematically predicted diffraction patterns of an hypothesized or "model" structure are compared to the actual pattern generated by the crystalline sample. Ideally, researchers make several initial guesses, which through refinement all converge on the same answer. Models are refined until their predicted patterns match to as great a degree as can be achieved without radical revision of the model. This is a painstaking process, made much easier today by computers.

The mathematical methods for the analysis of diffraction data only apply to patterns, which in turn result only when waves diffract from orderly arrays. Hence crystallography applies for the most part only to crystals, or to molecules which can be coaxed to crystallize for the sake of measurement. In spite of this, a certain amount of molecular information can be deduced from patterns that are generated by fibers and powders, which while not as perfect as a solid crystal, may exhibit a degree of order. This level of order can be sufficient to deduce the structure of simple molecules, or to determine the coarse features of more complicated molecules. For example, the double-helical structure of DNA was deduced from an X-ray diffraction pattern that had been generated by a fibrous sample.

Materials Science

Crystallography is used by materials scientists to characterize different materials. In single crystals, the effects of the crystalline arrangement of atoms is often easy to see macroscopically, because the natural shapes of crystals reflect the atomic structure. In addition, physical properties are often controlled by crystalline defects. The understanding of crystal structures is an important prerequisite for understanding crystallographic defects. Mostly, materials do not occur as a single crystal, but in poly-crystalline form (i.e., as an aggregate of small crystals with different orientations). Because of this, the powder diffraction method, which takes diffraction patterns of polycrystalline samples with a large number of crystals, plays an important role in structural determination.

Other physical properties are also linked to crystallography. For example, the minerals in clay form small, flat, platelike structures. Clay can be easily deformed because the platelike particles can slip along each other in the plane of the plates, yet remain strongly connected in the direction perpendicular to the plates. Such mechanisms can be studied by crystallographic texture measurements.

In another example, iron transforms from a body-centered cubic (bcc) structure to a face-centered cubic (fcc) structure called austenite when it is heated. The fcc structure is a close-packed structure unlike the bcc structure; thus the volume of the iron decreases when this transformation occurs.

Crystallography is useful in phase identification. When manufacturing or using a material, it is generally desirable to know what compounds and what phases are present in the material, as their composition, structure and proportions will influence the material's properties. Each phase has a characteristic arrangement of atoms. X-ray or neutron diffraction can be used to identify which patterns are present in the material, and thus which compounds are present. Crystallography covers the enumeration of the symmetry patterns which can be formed by atoms in a crystal and for this reason is related to group theory and geometry.

Biology

X-ray crystallography is the primary method for determining the molecular conformations of biological macromolecules, particularly protein and nucleic acids such as DNA and RNA. In fact, the double-helical structure of DNA was deduced from crystallographic data. The first crystal structure of a macromolecule was solved in 1958, a three-dimensional model of the myoglobin molecule obtained by X-ray analysis. The Protein Data Bank (PDB) is a freely accessible repository for the structures of proteins and other biological macromolecules. Computer programs such as RasMol, Pymol or VMD can be used to visualize biological molecular structures. Neutron crystallography is often used to help refine structures obtained by X-ray methods or to solve a specific bond; the methods are often viewed as complementary, as X-rays are sensitive to electron positions and scatter most strongly off heavy atoms, while neutrons are sensitive to nucleus positions and scatter strongly even off many light isotopes, including hydrogen and deuterium. Electron crystallography has been used to determine some protein structures, most notably membrane proteins and viral capsids.

Crystal

A crystal or crystalline solid is a solid material whose constituents (such as atoms, molecules, or ions) are arranged in a highly ordered microscopic structure, forming a crystal lattice that extends in all directions. In addition, macroscopic single crystals are usually identifiable by their geometrical shape, consisting of flat faces with specific, characteristic orientations. The scientific study of crystals and crystal formation is known as crystallography. The process of crystal formation via mechanisms of crystal growth is called crystallization or solidification.

Examples of large crystals include snowflakes, diamonds, and table salt. Most inorganic solids are not crystals but polycrystals, i.e. many microscopic crystals fused together into a single solid. Examples of polycrystals include most metals, rocks, ceramics, and ice. A third category of solids is amorphous solids, where the atoms have no periodic structure whatsoever. Examples of amorphous solids include glass, wax, and many plastics.

Despite the name, lead crystal, crystal glass, and related products are *not* crystals, but rather types of glass, i.e. amorphous solids.

Crystals are often used in pseudoscientific practices such as crystal therapy, and, along with gemstones, are sometimes associated with spellwork in Wiccan beliefs and related religious movements.

Crystal Structure (Microscopic)

The scientific definition of a "crystal" is based on the microscopic arrangement of atoms inside it, called the crystal structure. A crystal is a solid where the atoms form a periodic arrangement.

Not all solids are crystals. For example, when liquid water starts freezing, the phase change begins with small ice crystals that grow until they fuse, forming a polycrystalline structure. In the final block of ice, each of the small crystals (called "crystallites" or "grains") is a true crystal with a periodic arrangement of atoms, but the whole polycrystal does not have a periodic arrangement of atoms, because the periodic pattern is broken at the grain boundaries. Most macroscopic inorganic solids are polycrystalline, including almost all metals, ceramics, ice, rocks, etc. Solids that are neither crystalline nor polycrystalline, such as glass, are called amorphous solids, also called glassy, vitreous, or noncrystalline. These have no periodic order, even microscopically. There are distinct differences between crystalline solids and amorphous solids: most notably, the process of forming a glass does not release the latent heat of fusion, but forming a crystal does.

Halite (table salt, NaCl): Microscopic and macroscopic	
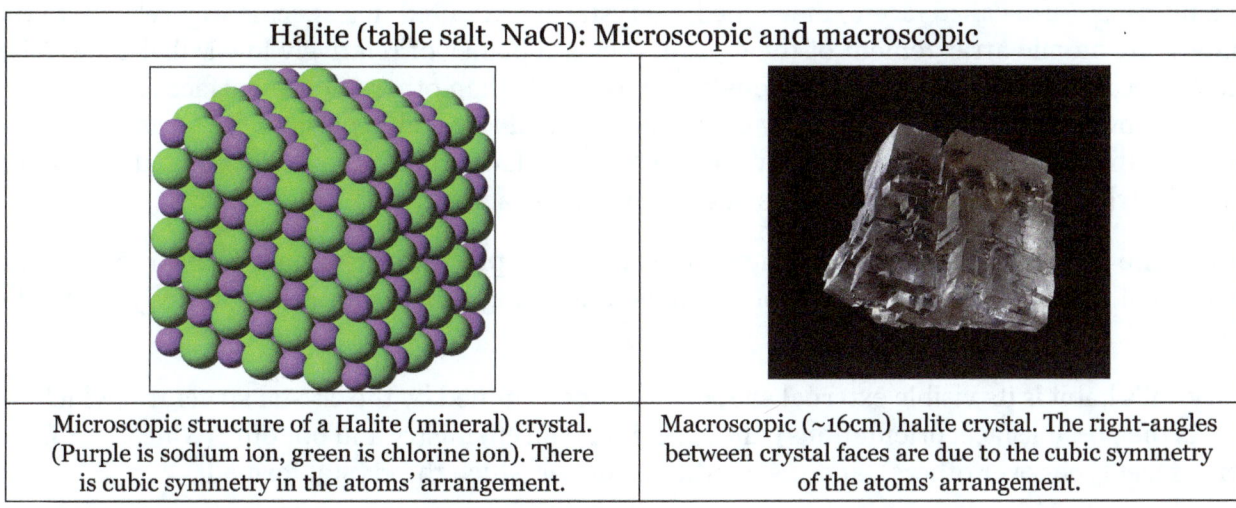	
Microscopic structure of a Halite (mineral) crystal. (Purple is sodium ion, green is chlorine ion). There is cubic symmetry in the atoms' arrangement.	Macroscopic (~16cm) halite crystal. The right-angles between crystal faces are due to the cubic symmetry of the atoms' arrangement.

A crystal structure (an arrangement of atoms in a crystal) is characterized by its *unit cell*, a small imaginary box containing one or more atoms in a specific spatial arrangement. The unit cells are stacked in three-dimensional space to form the crystal.

The symmetry of a crystal is constrained by the requirement that the unit cells stack perfectly with no gaps. There are 219 possible crystal symmetries, called crystallographic space groups. These are grouped into 7 crystal systems, such as cubic crystal system (where the crystals may form cubes or rectangular boxes, such as Halite (mineral) shown at above) or hexagonal crystal system (where the crystals may form hexagons, such as ordinary water ice).

Crystal Faces and Shapes

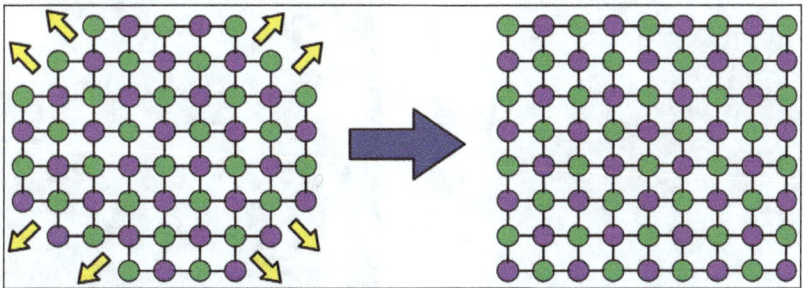

As a Halite (mineral) crystal is growing, new atoms can very easily attach to the parts of the surface with rough atomic-scale structure and many dangling bonds. Therefore, these parts of the crystal grow out very quickly (yellow arrows). Eventually, the whole surface consists of smooth, stable faces, where new atoms cannot as easily attach themselves.

Crystals are commonly recognized by their shape, consisting of flat faces with sharp angles. These shape characteristics are not necessary for a crystal—a crystal is scientifically defined by its microscopic atomic arrangement, not its macroscopic shape—but the characteristic macroscopic shape is often present and easy to see.

Euhedral crystals are those with obvious, well-formed flat faces. Anhedral crystals do not, usually because the crystal is one grain in a polycrystalline solid.

The flat faces (also called facets) of a euhedral crystal are oriented in a specific way relative to the underlying atomic arrangement of the crystal: they are planes of relatively low Miller index.This occurs because some surface orientations are more stable than others (lower surface energy). As a crystal grows, new atoms attach easily to the rougher and less stable parts of the surface, but less easily to the flat, stable surfaces. Therefore, the flat surfaces tend to grow larger and smoother, until the whole crystal surface consists of these plane surfaces.

One of the oldest techniques in the science of crystallography consists of measuring the three-dimensional orientations of the faces of a crystal, and using them to infer the underlying crystal symmetry.

A crystal's habit is its visible external shape. This is determined by the crystal structure (which restricts the possible facet orientations), the specific crystal chemistry and bonding (which may favor some facet types over others), and the conditions under which the crystal formed.

Occurrence in Nature

Rocks

By volume and weight, the largest concentrations of crystals in the Earth are part of its solid bedrock. Crystals found in rocks typically range in size from a fraction of a millimetre to several centimetres across, although exceptionally large crystals are occasionally found. As of 1999, the world's largest known naturally occurring crystal is a crystal of beryl from Malakialina, Madagascar, 18 m (59 ft) long and 3.5 m (11 ft) in diameter, and weighing 380,000 kg (840,000 lb).

Ice Crystals.

Fossil shell with calcite crystals.

Some crystals have formed by magmatic and metamorphic processes, giving origin to large masses of crystalline rock. The vast majority of igneous rocks are formed from molten magma and the degree of crystallization depends primarily on the conditions under which they solidified. Such rocks as granite, which have cooled very slowly and under great pressures, have completely crystallized; but many kinds of lava were poured out at the surface and cooled very rapidly, and in this latter group a small amount of amorphous or glassy matter is common. Other crystalline rocks, the metamorphic rocks such as marbles, mica-schists and quartzites, are recrystallized. This means that they were at first fragmental rocks like limestone, shale and sandstone and have never been in a molten condition nor entirely in solution, but the high temperature and pressure conditions of metamorphism have acted on them by erasing their original structures and inducing recrystallization in the solid state.

Other rock crystals have formed out of precipitation from fluids, commonly water, to form druses or quartz veins. Evaporites such as Halite (mineral), gypsum and some limestones have been deposited from aqueous solution, mostly owing to evaporation in arid climates.

Ice

Water-based ice in the form of snow, sea ice and glaciers is a very common manifestation of crystalline or polycrystalline matter on Earth. A single snowflake is a single crystal or a collection of crystals,while an ice cube is a polycrystal.

Organigenic Crystals

Many living organisms are able to produce crystals, for example calcite and aragonite in the case of most molluscs or hydroxylapatite in the case of vertebrates.

Polymorphism and Allotropy

The same group of atoms can often solidify in many different ways. Polymorphism is the ability of a solid to exist in more than one crystal form. For example, water ice is ordinarily found in the hexagonal form Ice I_h, but can also exist as the cubic Ice I_c, the rhombohedral ice II, and many other forms. The different polymorphs are usually called different *phases*.

In addition, the same atoms may be able to form noncrystalline phases. For example, water can also form amorphous ice, while SiO_2 can form both fused silica (an amorphous glass) and quartz (a crystal). Likewise, if a substance can form crystals, it can also form polycrystals.

For pure chemical elements, polymorphism is known as allotropy. For example, diamond and graphite are two crystalline forms of carbon, while amorphous carbon is a noncrystalline form. Polymorphs, despite having the same atoms, may have wildly different properties. For example, diamond is among the hardest substances known, while graphite is so soft that it is used as a lubricant.

Polyamorphism is a similar phenomenon where the same atoms can exist in more than one amorphous solid form.

Crystallization

Crystallization is the process of forming a crystalline structure from a fluid or from materials dissolved in a fluid. (More rarely, crystals may be deposited directly from gas; see thin-film deposition and epitaxy.)

Vertical cooling crystallizer in a beet sugar factory.

Crystallization is a complex and extensively-studied field, because depending on the conditions, a single fluid can solidify into many different possible forms. It can form a single crystal, perhaps with various possible phases, stoichiometries, impurities, defects, and habits. Or, it can form a polycrystal, with various possibilities for the size, arrangement, orientation, and phase of its grains. The final form of the solid is determined by the conditions under which the fluid is being solidified, such as the chemistry of the fluid, the ambient pressure, the temperature, and the speed with which all these parameters are changing.

Specific industrial techniques to produce large single crystals (called *boules*) include the Czochralski process and the Bridgman technique. Other less exotic methods of crystallization may be used, depending on the physical properties of the substance, including hydrothermal synthesis, sublimation, or simply solvent-based crystallization.

Large single crystals can be created by geological processes. For example, selenite crystals in excess of 10 meters are found in the Cave of the Crystals in Naica, Mexico.

Crystals can also be formed by biological processes. Conversely, some organisms have special techniques to prevent crystallization from occurring, such as antifreeze proteins.

Defects, Impurities and Twinning

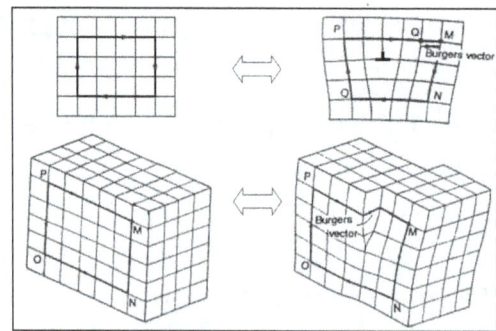

Two types of crystallographic defects. Top right: edge dislocation.
Bottom right: screw dislocation.

An ideal crystal has every atom in a perfect, exactly repeating pattern. However, in reality, most crystalline materials have a variety of crystallographic defects, places where the crystal's pattern is interrupted. The types and structures of these defects may have a profound effect on the properties of the materials.

A few examples of crystallographic defects include vacancy defects (an empty space where an atom should fit), interstitial defects (an extra atom squeezed in where it does not fit), and dislocations. Dislocations are especially important in materials science, because they help determine the mechanical strength of materials.

Another common type of crystallographic defect is an impurity, meaning that the "wrong" type of atom is present in a crystal. For example, a perfect crystal of diamond would only contain carbon atoms, but a real crystal might perhaps contain a few boron atoms as well. These boron impurities change the diamond's color to slightly blue. Likewise, the only difference between ruby and sapphire is the type of impurities present in a corundum crystal.

Twinned pyrite crystal group.

In semiconductors, a special type of impurity, called a dopant, drastically changes the crystal's electrical properties. Semiconductor devices, such as transistors, are made possible largely by putting different semiconductor dopants into different places, in specific patterns.

Twinning is a phenomenon somewhere between a crystallographic defect and a grain boundary. Like a grain boundary, a twin boundary has different crystal orientations on its two sides. But unlike a grain boundary, the orientations are not random, but related in a specific, mirror-image way.

Mosaicity is a spread of crystal plane orientations. A mosaic crystal is supposed to consist of smaller crystalline units that are somewhat misaligned with respect to each other.

Chemical Bonds

In general, solids can be held together by various types of chemical bonds, such as metallic bonds, ionic bonds, covalent bonds, van der Waals bonds, and others. None of these are necessarily crystalline or non-crystalline. However, there are some general trends as follows.

Metals are almost always polycrystalline, though there are exceptions like amorphous metal and single-crystal metals. The latter are grown synthetically. (A microscopically-small piece of metal may naturally form into a single crystal, but larger pieces generally do not.) Ionic compound materials are usually crystalline or polycrystalline. In practice, large salt crystals can be created by solidification of a molten fluid, or by crystallization out of a solution. Covalently bonded solids (sometimes called covalent network solids) are also very common, notable examples being diamond and quartz. Weak van der Waals forces also help hold together certain crystals, such as crystalline molecular solids, as well as the interlayer bonding in graphite. Polymer materials generally will form crystalline regions, but the lengths of the molecules usually prevent complete crystallization—and sometimes polymers are completely amorphous.

Quasicrystals

The material holmium–magnesium–zinc (Ho–Mg–Zn) forms quasicrystals, which can take on the macroscopic shape of a dodecahedron. (Only a quasicrystal, not a normal crystal, can take this shape.) The edges are 2 mm long.

A quasicrystal consists of arrays of atoms that are ordered but not strictly periodic. They have many attributes in common with ordinary crystals, such as displaying a discrete pattern in x-ray diffraction, and the ability to form shapes with smooth, flat faces.

Quasicrystals are most famous for their ability to show five-fold symmetry, which is impossible for an ordinary periodic crystal.

The International Union of Crystallography has redefined the term "crystal" to include both ordinary periodic crystals and quasicrystals ("any solid having an essentially discrete diffraction diagram").

Quasicrystals, first discovered in 1982, are quite rare in practice. Only about 100 solids are known to form quasicrystals, compared to about 400,000 periodic crystals known in 2004.

Special Properties from Anisotropy

Crystals can have certain special electrical, optical, and mechanical properties that glass and polycrystals normally cannot. These properties are related to the anisotropy of the crystal, i.e. the lack of rotational symmetry in its atomic arrangement. One such property is the piezoelectric effect, where a voltage across the crystal can shrink or stretch it. Another is birefringence, where a double image appears when looking through a crystal. Moreover, various properties of a crystal, including electrical conductivity, electrical permittivity, and Young's modulus, may be different in different directions in a crystal. For example, graphite crystals consist of a stack of sheets, and although each individual sheet is mechanically very strong, the sheets are rather loosely bound to each other. Therefore, the mechanical strength of the material is quite different depending on the direction of stress.

Not all crystals have all of these properties. Conversely, these properties are not quite exclusive to crystals. They can appear in glasses or polycrystals that have been made anisotropic by working or stress—for example, stress-induced birefringence.

Crystallography

Crystallography is the science of measuring the crystal structure (in other words, the atomic arrangement) of a crystal. One widely used crystallography technique is X-ray diffraction. Large numbers of known crystal structures are stored in crystallographic databases.

Crystal Structure

In crystallography, crystal structure is a description of the ordered arrangement of atoms, ions or molecules in a crystalline material. Ordered structures occur from the intrinsic nature of the constituent particles to form symmetric patterns that repeat along the principal directions of three-dimensional space in matter.

The smallest group of particles in the material that constitutes this repeating pattern is the unit cell of the structure. The unit cell completely reflects the symmetry and structure of the entire crystal, which is built up by repetitive translation of the unit cell along its principal axes. The translation vectors define the nodes of the Bravais lattice.

The lengths of the principal axes, or edges, of the unit cell and the angles between them are the lattice constants, also called lattice parameters or cell parameters. The symmetry properties of the crystal are described by the concept of space groups. All possible symmetric arrangements of particles in three-dimensional space may be described by the 230 space groups.

The crystal structure and symmetry play a critical role in determining many physical properties, such as cleavage, electronic band structure, and optical transparency.

Unit Cell

Crystal structure is described in terms of the geometry of arrangement of particles in the unit cell. The unit cell is defined as the smallest repeating unit having the full symmetry of the crystal structure. The geometry of the unit cell is defined as a parallelepiped, providing six lattice parameters taken as the lengths of the cell edges (a, b, c) and the angles between them (α, β, γ). The positions of particles inside the unit cell are described by the fractional coordinates (x_i, y_i, z_i) along the cell edges, measured from a reference point. It is only necessary to report the coordinates of a smallest asymmetric subset of particles. This group of particles may be chosen so that it occupies the smallest physical space, which means that not all particles need to be physically located inside the boundaries given by the lattice parameters. All other particles of the unit cell are generated by the symmetry operations that characterize the symmetry of the unit cell. The collection of symmetry operations of the unit cell is expressed formally as the space group of the crystal structure.

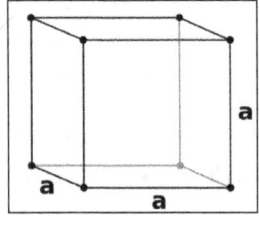

Simple cubic (P)

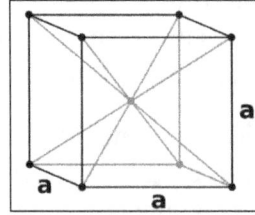

Body-centered cubic (I)

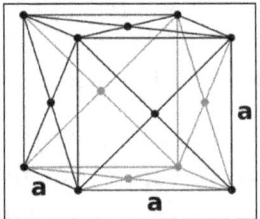

Face-centered cubic (F)

Miller Indices

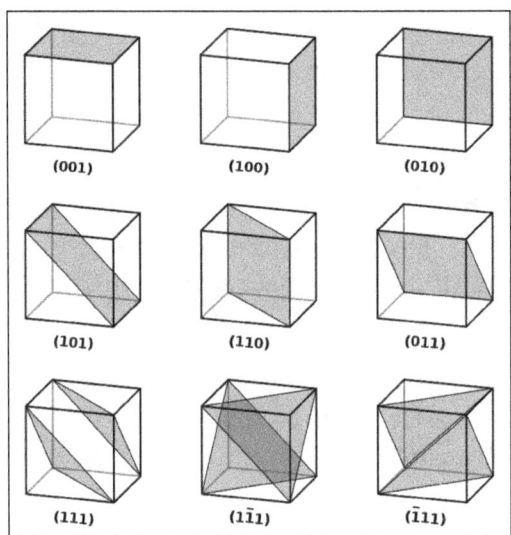

Planes with different Miller indices in cubic crystals.

Vectors and planes in a crystal lattice are described by the three-value Miller index notation. This syntax uses the indices ℓ, m, and n as directional orthogonal parameters, which are separated by $90°$.

By definition, the syntax (ℓmn) denotes a plane that intercepts the three points a_1/ℓ, a_2/m, and a_3/n, or some multiple thereof. That is, the Miller indices are proportional to the inverses of the intercepts of the plane with the unit cell (in the basis of the lattice vectors). If one or more of the indices is zero, it means that the planes do not intersect that axis (i.e., the intercept is "at infinity"). A plane containing a coordinate axis is translated so that it no longer contains that axis before its Miller indices are determined. The Miller indices for a plane are integers with no common factors. Negative indices are indicated with horizontal bars, as in $(1\bar{2}3)$. In an orthogonal coordinate system for a cubic cell, the Miller indices of a plane are the Cartesian components of a vector normal to the plane.

Considering only (ℓmn) planes intersecting one or more lattice points (the lattice planes), the distance d between adjacent lattice planes is related to the (shortest) reciprocal lattice vector orthogonal to the planes by the formula:

$$d = \frac{2\pi}{|g_{\ell mn}|}$$

Planes and Directions

The crystallographic directions are geometric lines linking nodes (atoms, ions or molecules) of a crystal. Likewise, the crystallographic planes are geometric planes linking nodes. Some directions and planes have a higher density of nodes. These high density planes have an influence on the behavior of the crystal as follows:

- Optical properties: Refractive index is directly related to density (or periodic density fluctuations).

- Adsorption and reactivity: Physical adsorption and chemical reactions occur at or near surface atoms or molecules. These phenomena are thus sensitive to the density of nodes.

- Surface tension: The condensation of a material means that the atoms, ions or molecules are more stable if they are surrounded by other similar species. The surface tension of an interface thus varies according to the density on the surface.

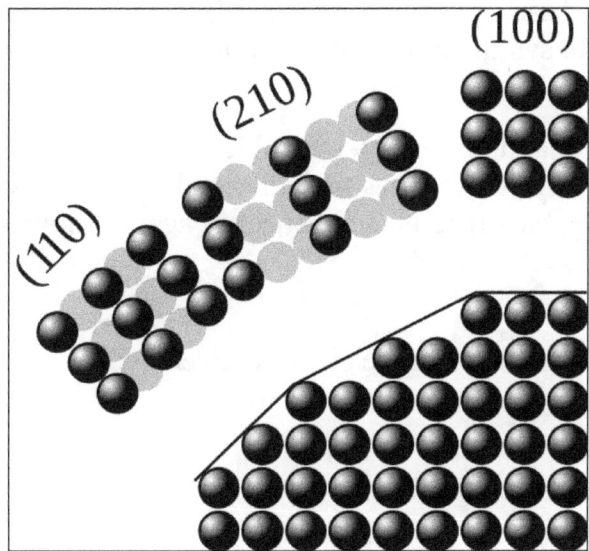
Dense crystallographic planes.

- Microstructural defects: Pores and crystallites tend to have straight grain boundaries following higher density planes.

- Cleavage: This typically occurs preferentially parallel to higher density planes.

- Plastic deformation: Dislocation glide occurs preferentially parallel to higher density planes. The perturbation carried by the dislocation (Burgers vector) is along a dense direction. The shift of one node in a more dense direction requires a lesser distortion of the crystal lattice.

Some directions and planes are defined by symmetry of the crystal system. In monoclinic, rhombohedral, tetragonal, and trigonal/hexagonal systems there is one unique axis (sometimes called the principal axis) which has higher rotational symmetry than the other two axes. The basal plane is the plane perpendicular to the principal axis in these crystal systems. For triclinic, orthorhombic, and cubic crystal systems the axis designation is arbitrary and there is no principal axis.

Cubic Structures

For the special case of simple cubic crystals, the lattice vectors are orthogonal and of equal length (usually denoted a); similarly for the reciprocal lattice. So, in this common case, the Miller indices (ℓmn) and $[\ell mn]$ both simply denote normals/directions in Cartesian coordinates. For cubic crystals with lattice constant a, the spacing d between adjacent (ℓmn) lattice planes is (from above):

$$d_{\ell mn} = \frac{a}{\sqrt{\ell^2 + m^2 + n^2}}$$

Because of the symmetry of cubic crystals, it is possible to change the place and sign of the integers and have equivalent directions and planes:

- Coordinates in angle brackets such as ⟨100⟩ denote a family of directions that are equivalent due to symmetry operations, such as [100], [010], or the negative of any of those directions.

- Coordinates in curly brackets or braces such as {100} denote a family of plane normals that are equivalent due to symmetry operations, much the way angle brackets denote a family of directions.

For face-centered cubic (fcc) and body-centered cubic (bcc) lattices, the primitive lattice vectors are not orthogonal. However, in these cases the Miller indices are conventionally defined relative to the lattice vectors of the cubic supercell and hence are again simply the Cartesian directions.

Interplanar Spacing

The spacing d between adjacent (hkl) lattice planes is given by:

- Cubic:

$$\frac{1}{d^2} = \frac{h^2 + k^2 + l^2}{a^2}$$

- Tetragonal:

$$\frac{1}{d^2} = \frac{h^2 + k^2}{a^2} + \frac{l^2}{c^2}$$

- Hexagonal:

$$\frac{1}{d^2} = \frac{4}{3}\left(\frac{h^2 + hk + k^2}{a^2}\right) + \frac{l^2}{c^2}$$

- Rhombohedral:

$$\frac{1}{d^2} = \frac{(h^2 + k^2 + l^2)\sin^2\alpha + 2(hk + kl + hl)(\cos^2\alpha - \cos\alpha)}{a^2(1 - 3\cos^2\alpha + 2\cos^3\alpha)}$$

- Orthorhombic:

$$\frac{1}{d^2} = \frac{h^2}{a^2} + \frac{k^2}{b^2} + \frac{l^2}{c^2}$$

- Monoclinic:

$$\frac{1}{d^2} = \left(\frac{h^2}{a^2} + \frac{k^2 \sin^2 \beta}{b^2} + \frac{l^2}{c^2} - \frac{2hl \cos \beta}{ac} \right) \cos^2 \beta$$

- Triclinic:

$$\frac{1}{d^2} = \frac{\dfrac{h^2}{a^2}\sin^2\alpha + \dfrac{k^2}{b^2}\sin^2\beta + \dfrac{l^2}{c^2}\sin^2\gamma + \dfrac{2kl}{bc}(\cos\beta\cos\gamma - \cos\alpha) + \dfrac{2hl}{ac}(\cos\gamma\cos\alpha - \cos\beta) + \dfrac{2hk}{ab}(\cos\alpha\cos\beta - \cos\gamma)}{1 - \cos^2\alpha - \cos^2\beta - \cos^2\gamma + 2\cos\alpha\cos\beta\cos\gamma}$$

Classification by Symmetry

The defining property of a crystal is its inherent symmetry. Performing certain symmetry operations on the crystal lattice leaves it unchanged. All crystals have translational symmetry in three directions, but some have other symmetry elements as well. For example, rotating the crystal 180° about a certain axis may result in an atomic configuration that is identical to the original configuration; the crystal has twofold rotational symmetry about this axis. In addition to rotational symmetry, a crystal may have symmetry in the form of mirror planes, and also the so-called compound symmetries, which are a combination of translation and rotation or mirror symmetries. A full classification of a crystal is achieved when all inherent symmetries of the crystal are identified.

Lattice Systems

Lattice systems are a grouping of crystal structures according to the axial system used to describe their lattice. Each lattice system consists of a set of three axes in a particular geometric arrangement. All crystals fall into one of seven lattice systems. They are similar to, but not quite the same as the seven crystal systems.

The simplest and most symmetric, the cubic or isometric system, has the symmetry of a cube, that is, it exhibits four threefold rotational axes oriented at 109.5° (the tetrahedral angle) with respect to each other. These threefold axes lie along the body diagonals of the cube. The other six lattice systems, are hexagonal, tetragonal, rhombohedral (often confused with the trigonal crystal system), orthorhombic, monoclinic and triclinic.

Bravais Lattices

Bravais lattices, also referred to as space lattices, describe the geometric arrangement of the lattice points, and therefore the translational symmetry of the crystal. The three dimensions of space afford 14 distinct Bravais lattices describing the translational symmetry. All crystalline materials recognized today, not including quasicrystals, fit in one of these arrangements.

The crystal structure consists of the same group of atoms, the basis, positioned around each and every lattice point. This group of atoms therefore repeats indefinitely in three dimensions according to the arrangement of one of the Bravais lattices. The characteristic rotation and mirror symmetries of the unit cell is described by its crystallographic point group.

Crystal Systems

A crystal system is a set of point groups in which the point groups themselves and their corresponding space groups are assigned to a lattice system. Of the 32 point groups that exist in three dimensions, most are assigned to only one lattice system, in which case the crystal system and lattice system both have the same name. However, five point groups are assigned to two lattice systems, rhombohedral and hexagonal, because both lattice systems exhibit threefold rotational symmetry. These point groups are assigned to the trigonal crystal system.

Point Groups

The crystallographic point group or *crystal class* is the mathematical group comprising the symmetry operations that leave at least one point unmoved and that leave the appearance of the crystal structure unchanged. These symmetry operations include:

- Reflection which reflects the structure across a reflection plane.

- Rotation which rotates the structure a specified portion of a circle about a rotation axis.

- Inversion which changes the sign of the coordinate of each point with respect to a center of symmetry or inversion point.

- Improper rotation which consists of a rotation about an axis followed by an inversion.

Rotation axes (proper and improper), reflection planes, and centers of symmetry are collectively called *symmetry elements*. There are 32 possible crystal classes. Each one can be classified into one of the seven crystal systems.

Space Groups

In addition to the operations of the point group, the space group of the crystal structure contains translational symmetry operations. These include:

- Pure translations, which move a point along a vector.

- Screw axes, which rotate a point around an axis while translating parallel to the axis.

- Glide planes, which reflect a point through a plane while translating it parallel to the plane.

There are 230 distinct space groups.

Atomic Coordination

By considering the arrangement of atoms relative to each other, their coordination numbers (or number of nearest neighbors), interatomic distances, types of bonding, etc., it is possible to form a general view of the structures and alternative ways of visualizing them.

Close Packing

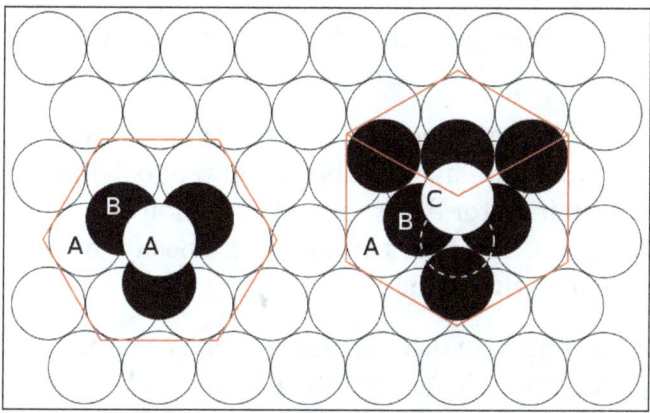

The hcp lattice (left) and the fcc lattice (right).

The principles involved can be understood by considering the most efficient way of packing together equal-sized spheres and stacking close-packed atomic planes in three dimensions. For example, if plane A lies beneath plane B, there are two possible ways of placing an additional atom on top of layer B. If an additional layer was placed directly over plane A, this would give rise to the following series:

...ABABABAB...

This arrangement of atoms in a crystal structure is known as hexagonal close packing (hcp).

If, however, all three planes are staggered relative to each other and it is not until the fourth layer is positioned directly over plane A that the sequence is repeated, then the following sequence arises:

...ABCABCABC...

This type of structural arrangement is known as cubic close packing (ccp).

The unit cell of a ccp arrangement of atoms is the face-centered cubic (fcc) unit cell. This is not immediately obvious as the closely packed layers are parallel to the {111} planes of the fcc unit cell. There are four different orientations of the close-packed layers.

The packing efficiency can be worked out by calculating the total volume of the spheres and dividing by the volume of the cell as follows:

$$\frac{4 \times \frac{4}{3}\pi r^3}{16\sqrt{2}r^3} = \frac{\pi}{3\sqrt{2}} = 0.7405$$

The 74% packing efficiency is the maximum density possible in unit cells constructed of spheres of only one size. Most crystalline forms of metallic elements are hcp, fcc, or bcc (body-centered cubic). The coordination number of atoms in hcp and fcc structures is 12 and its atomic packing factor (APF) is the number mentioned above, 0.74. This can be compared to the APF of a bcc structure, which is 0.68.

Grain Boundaries

Grain boundaries are interfaces where crystals of different orientations meet. A grain boundary is a single-phase interface, with crystals on each side of the boundary being identical except in orientation. The term "crystallite boundary" is sometimes, though rarely, used. Grain boundary areas contain those atoms that have been perturbed from their original lattice sites, dislocations, and impurities that have migrated to the lower energy grain boundary.

Treating a grain boundary geometrically as an interface of a single crystal cut into two parts, one of which is rotated, we see that there are five variables required to define a grain boundary. The first two numbers come from the unit vector that specifies a rotation axis. The third number designates the angle of rotation of the grain. The final two numbers specify the plane of the grain boundary (or a unit vector that is normal to this plane).

Grain boundaries disrupt the motion of dislocations through a material, so reducing crystallite size is a common way to improve strength, as described by the Hall–Petch relationship. Since grain boundaries are defects in the crystal structure they tend to decrease the electrical and thermal conductivity of the material. The high interfacial energy and relatively weak bonding in most grain boundaries often makes them preferred sites for the onset of corrosion and for the precipitation of new phases from the solid. They are also important to many of the mechanisms of creep.

Grain boundaries are in general only a few nanometers wide. In common materials, crystallites are large enough that grain boundaries account for a small fraction of the material. However, very small grain sizes are achievable. In nanocrystalline solids, grain boundaries become a significant volume fraction of the material, with profound effects on such properties as diffusion and plasticity. In the limit of small crystallites, as the volume fraction of grain boundaries approaches 100%, the material ceases to have any crystalline character, and thus becomes an amorphous solid.

Defects and Impurities

Real crystals feature defects or irregularities in the ideal arrangements described above and it is these defects that critically determine many of the electrical and mechanical properties of real materials. When one atom substitutes for one of the principal atomic components within the crystal structure, alteration in the electrical and thermal properties of the material may ensue. Impurities may also manifest as electron spin impurities in certain materials. Research on magnetic impurities demonstrates that substantial alteration of certain properties such as specific heat may be affected by small concentrations of an impurity, as for example impurities in semiconducting ferromagnetic alloys may lead to different properties as first predicted in the late 1960s. Dislocations in the crystal lattice allow shear at lower stress than that needed for a perfect crystal structure.

Prediction of Structure

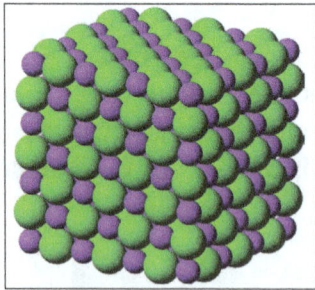

Crystal structure of sodium chloride (table salt).

The difficulty of predicting stable crystal structures based on the knowledge of only the chemical composition has long been a stumbling block on the way to fully computational materials design. Now, with more powerful algorithms and high-performance computing, structures of medium complexity can be predicted using such approaches as evolutionary algorithms, random sampling, or metadynamics.

The crystal structures of simple ionic solids (e.g., NaCl or table salt) have long been rationalized in terms of Pauling's rules, first set out in 1929 by Linus Pauling, referred to by many since as the "father of the chemical bond". Pauling also considered the nature of the interatomic forces in metals, and concluded that about half of the five d-orbitals in the transition metals are involved in bonding, with the remaining nonbonding d-orbitals being responsible for the magnetic properties. He, therefore, was able to correlate the number of d-orbitals in bond formation with the bond length as well as many of the physical properties of the substance. He subsequently introduced the metallic orbital, an extra orbital necessary to permit uninhibited resonance of valence bonds among various electronic structures.

In the resonating valence bond theory, the factors that determine the choice of one from among alternative crystal structures of a metal or intermetallic compound revolve around the energy of resonance of bonds among interatomic positions. It is clear that some modes of resonance would make larger contributions (be more mechanically stable than others), and that in particular a simple ratio of number of bonds to number of positions would be exceptional. The resulting principle is that a special stability is associated with the simplest ratios or "bond numbers": $\frac{1}{2}$, $\frac{1}{3}$, $\frac{2}{3}$, $\frac{1}{4}$, $\frac{3}{4}$, etc. The choice of structure and the value of the axial ratio (which determines the relative bond lengths) are thus a result of the effort of an atom to use its valency in the formation of stable bonds with simple fractional bond numbers.

After postulating a direct correlation between electron concentration and crystal structure in beta-phase alloys, Hume-Rothery analyzed the trends in melting points, compressibilities and bond lengths as a function of group number in the periodic table in order to establish a system of valencies of the transition elements in the metallic state. This treatment thus emphasized the increasing bond strength as a function of group number. The operation of directional forces were emphasized in one article on the relation between bond hybrids and the metallic structures. The resulting correlation between electronic and crystalline structures is summarized by a single parameter, the weight of the d-electrons per hybridized metallic orbital. The "d-weight" calculates out to 0.5, 0.7 and 0.9 for the fcc, hcp and bcc structures respectively. The relationship between d-electrons and crystal structure thus becomes apparent.

In crystal structure predictions/simulations, the periodicity is usually applied, since the system is imagined as unlimited big in all directions. Starting from a triclinic structure with no further symmetry property assumed, the system may be driven to show some additional symmetry properties by applying Newton's Second Law on particles in the unit cell and a recently developed dynamical equation for the system period vectors (lattice parameters including angles), even if the system is subject to external stress.

Polymorphism

Polymorphism is the occurrence of multiple crystalline forms of a material. It is found in many crystalline materials including polymers, minerals, and metals. According to Gibbs' rules of phase equilibria, these unique crystalline phases are dependent on intensive variables such as pressure and temperature. Polymorphism is related to allotropy, which refers to elemental solids. The complete morphology of a material is described by polymorphism and other variables such as crystal habit, amorphous fraction or crystallographic defects. Polymorphs have different stabilities and may spontaneously convert from a metastable form (or thermodynamically unstable form) to the stable form at a particular temperature. They also exhibit different melting points, solubilities, and X-ray diffraction patterns.

Quartz is one of the several crystalline forms of silica, SiO_2. The most important forms of silica include: α-quartz, β-quartz, tridymite, cristobalite, coesite, and stishovite.

One good example of this is the quartz form of silicon dioxide, or SiO_2. In the vast majority of silicates, the Si atom shows tetrahedral coordination by 4 oxygens. All but one of the crystalline forms involve tetrahedral $\{SiO_4\}$ units linked together by shared vertices in different arrangements. In different minerals the tetrahedra show different degrees of networking and polymerization. For example, they occur singly, joined together in pairs, in larger finite clusters including rings, in chains, double chains, sheets, and three-dimensional frameworks. The minerals are classified into groups based on these structures. In each of the 7 thermodynamically stable crystalline forms or polymorphs of crystalline quartz, only 2 out of 4 of each the edges of the $\{SiO_4\}$ tetrahedra are shared with others, yielding the net chemical formula for silica: SiO_2.

Another example is elemental tin (Sn), which is malleable near ambient temperatures but is brittle when cooled. This change in mechanical properties due to existence of its two major allotropes, α- and β-tin. The two allotropes that are encountered at normal pressure and temperature, α-tin and β-tin, are more commonly known as gray tin and white tin respectively. Two more allotropes, γ and σ, exist at temperatures above 161 °C and pressures above several GPa. White tin is metallic, and is the stable crystalline form at or above room temperature. Below 13.2 °C, tin exists in the gray form, which has a diamond cubic crystal structure, similar to diamond, silicon or germanium. Gray tin has no metallic properties at all, is a dull gray powdery material, and has few uses, other than a few specialized semiconductor applications. Although the α–β transformation temperature of tin is nominally 13.2 °C, impurities (e.g. Al, Zn, etc.) lower the transition temperature well below 0 °C, and upon addition of Sb or Bi the transformation may not occur at all.

Physical Properties

Twenty of the 32 crystal classes are piezoelectric, and crystals belonging to one of these classes (point groups) display piezoelectricity. All piezoelectric classes lack a center of symmetry. Any material develops a dielectric polarization when an electric field is applied, but a substance that has such a natural charge separation even in the absence of a field is called a polar material. Whether or not a material is polar is determined solely by its crystal structure. Only 10 of the 32 point groups are polar. All polar crystals are pyroelectric, so the 10 polar crystal classes are sometimes referred to as the pyroelectric classes.

There are a few crystal structures, notably the perovskite structure, which exhibit ferroelectric behavior. This is analogous to ferromagnetism, in that, in the absence of an electric field during production, the ferroelectric crystal does not exhibit a polarization. Upon the application of an electric field of sufficient magnitude, the crystal becomes permanently polarized. This polarization can be reversed by a sufficiently large counter-charge, in the same way that a ferromagnet can be reversed. However, although they are called ferroelectrics, the effect is due to the crystal structure (not the presence of a ferrous metal).

Crystal Forms

A crystal form is a set of crystal faces that are related to each other by symmetry. To designate a crystal form (which could imply many faces) we use the Miller Index, or Miller-Bravais Index notation enclosing the indices in curly braces, such as given below.

$\{101\}$ or $\{11\bar{2}1\}$

Such notation is called a form symbol.

An important point to note is that a form refers to a face or set of faces that have the same arrangement of atoms. Thus, the number of faces in a form depends on the symmetry of the crystal.

General Forms and Special Forms

A general form is a form in a particular crystal class that contains faces that intersect all crystallographic axes at different lengths. It has the form symbol $\{hkl\}$ All other forms that may be present are called special forms. In the monoclinic, triclinic, and orthorhombic crystal systems, the form $\{111\}$ is a general form because in these systems faces of this form will intersect the a, b, and c axes at different lengths because the unit lengths are different on each axis. In crystals of higher symmetry, where two or more of the axes have equal length, a general form must intersect the equal length axes at different multiples of the unit length. Thus in the tetragonal system the form $\{121\}$ is a general form. In the isometric system a general form would have to be something like $\{123\}$.

Open Forms and Closed Forms

A closed form is a set of crystal faces that completely enclose space. Thus, in crystal classes that contain closed forms, a crystal can be made up of a single form.

An open form is one or more crystal faces that do not completely enclose space.

Example 1. Pedions are single faced forms. Since there is only one face in the form a pedion cannot completely enclose space. Thus, a crystal that has only pedions, must have at least 3 different pedions to completely enclose space.

Example 2. A prism is a 3 or more faced form wherein the crystal faces are all parallel to the same line. If the faces are all parallel then they cannot completely enclose space. Thus crystals that have prisms must also have at least one additional form in order to completely enclose space.

Example 3. A dipyramid has at least 6 faces that meet in points at opposite ends of the crystal. These faces can completely enclose space, so a dipyramid is closed form. Although a crystal may be made up of a single dipyramid form, it may also have other forms present.

There are 48 possible forms that can be developed as the result of the 32 combinations of symmetry. We here discuss some, but not all of these forms.

Pedions

A pedion is an open, one faced form. Pedions are the only forms that occur in the Pedial class (1). Since a pedion is not related to any other face by symmetry, each form symbol refers to a single face. For example the form {100} refers only to the face (100), and is different from the form {00} which refers only to the face (00). Note that while forms in the Pedial class are pedions, pedions may occur in other crystal classes.

Pinacoids

A Pinacoid is an open 2-faced form made up of two parallel faces. In the crystal drawing shown here the form {111} is a pinacoid and consists of two faces, (111) and ($\bar{1}\bar{1}\bar{1}$). The form {100} is also a pinacoid consisting of the two faces (100) and (00). Similarly the form {010} is a pinacoid consisting of the two faces (010) and (00), and the form {001} is a two faced form consisting of the faces (001) and (00). In this case, note that at least three of the above forms are necessary to completely enclose space. While all forms in the Pinacoid class are pinacoids, pinacoids may occur in other crystal classes as well.

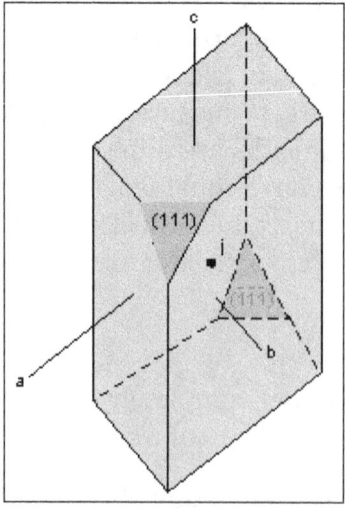

Domes

Domes are 2-faced open forms where the 2 faces are related to one another by a mirror plane. In the crystal model shown here, the dark shaded faces belong to a dome. The vertical faces along the side of the model are pinacoids (2 parallel faces). The faces on the front and back of the model are not related to each other by symmetry, and are thus two different pedions.

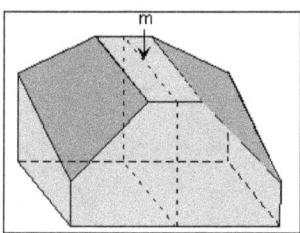

Sphenoids

Sphenoids are 2-faced open forms where the faces are related to each other by a 2-fold rotation axis and are not parallel to each other. The dark shaded triangular faces on the model shown here belong to a sphenoid. Pairs of similar vertical faces that cut the edges of the drawing are also pinacoids. The top and bottom faces, however, are two different pedions.

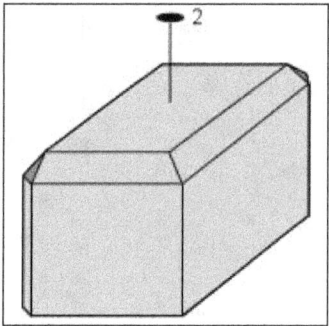

Prisms

A prism is an open form consisting of three or more parallel faces. Depending on the symmetry, several different kinds of prisms are possible.

- Trigonal prism: 3-faced form with all faces parallel to a 3-fold rotation axis.

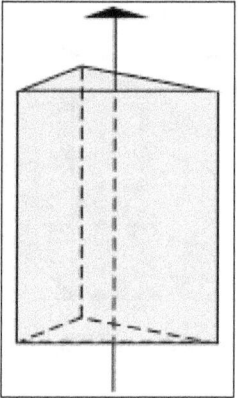

- Ditrigonal prism: 6-faced form with all 6 faces parallel to a 3-fold rotation axis. Note that the cross section of this form (shown to the right of the drawing) is not a hexagon, i.e. it does not have 6-fold rotational symmetry.

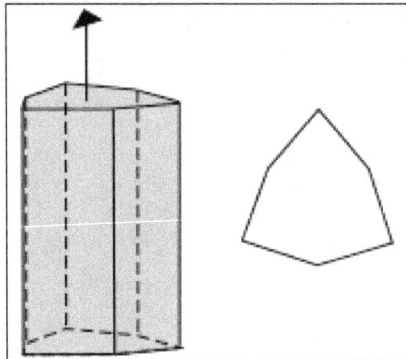

- Rhombic prism: 4-faced form with all faces parallel to a line that is not a symmetry element. In the drawing to the right, the 4 shaded faces belong to a rhombic prism. The other faces in this model are pinacoids (the faces on the sides belong to a side pinacoid, and the faces on the top and bottom belong to a top/bottom pinacoid).

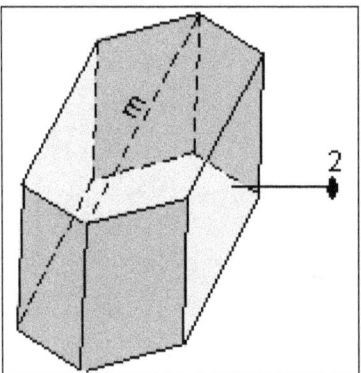

- Tetragonal prism: 4-faced open form with all faces parallel to a 4-fold rotation axis or $\bar{4}$. The 4 side faces in this model make up the tetragonal prism. The top and bottom faces make up the a form called the top/bottom pinacoid.

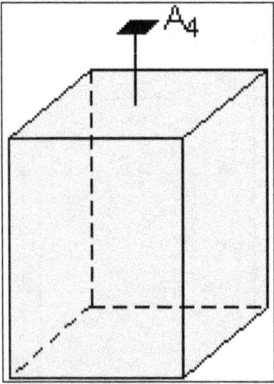

- Ditetragonal prism: 8-faced form with all faces parallel to a 4-fold rotation axis. In the drawing, the 8 vertical faces make up the ditetragonal prism.

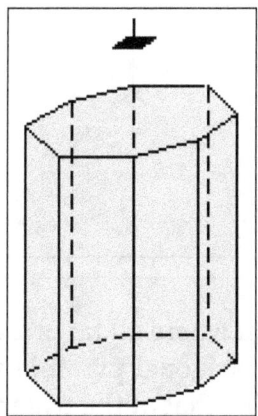

- Hexagonal prism: 6-faced form with all faces parallel to a 6-fold rotation axis. The 6 vertical faces in the drawing make up the hexagonal prism. Again the faces on top and bottom are the top/bottom pinacoid form.

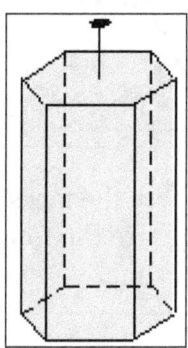

- Dihexagonal prism: 12-faced form with all faces parallel to a 6-fold rotation axis. Note that a horizontal cross-section of this model would have apparent 12-fold rotation symmetry. The dihexagonal prism is the result of mirror planes parallel to the 6-fold rotation axis.

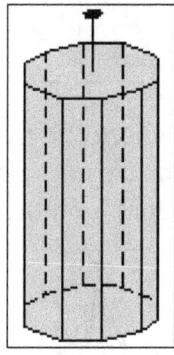

Pyramids

A pyramid is a 3, 4, 6, 8 or 12 faced open form where all faces in the form meet, or could meet if extended, at a point.

- Trigonal pyramid: 3-faced form where all faces are related by a 3-fold rotation axis.

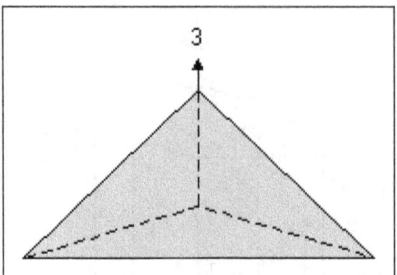

- Ditrigonal pyramid: 6-faced form where all faces are related by a 3-fold rotation axis. Note that if viewed from above, the ditrigonal pyramid would not have a hexagonal shape; its cross section would look more like that of the trigonal prism.

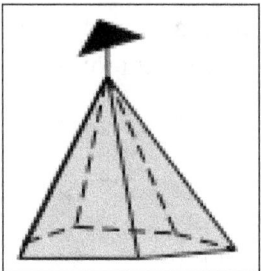

- Rhombic pyramid: 4-faced form where the faces are related by mirror planes. In the drawing shown here the faces labeled "p" are the four faces of the rhombic pyramid. If extend, these 4 faces would meet at a point.

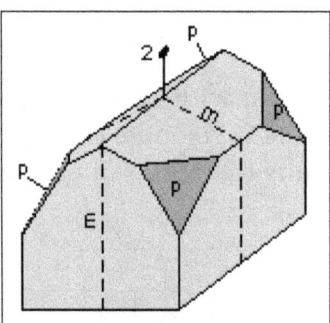

- Tetragonal pyramid: 4-faced form where the faces are related by a 4 axis. In the drawing the small triangular faces that cut the corners represent the tetragonal pyramid. Note that if extended, these 4 faces would meet at a point.

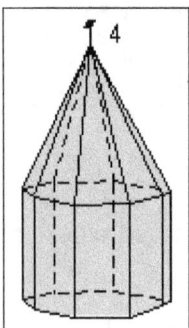

- Ditetragonal pyramid: 8-faced form where all faces are related by a 4 axis. In the drawing shown here, the upper 8 faces belong to the ditetragonal pyramid form. Note that the vertical faces belong to the ditetragonal prism.

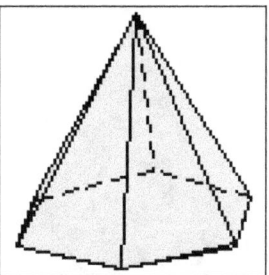

- Hexagonal pyramid: 6-faced form where all faces are related by a 6 axis. If viewed from above, the hexagonal pyramid would have a hexagonal shape.

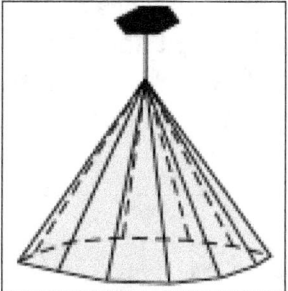

- Dihexagonal pyramid: 12-faced form where all faces are related by a 6-fold axis. This form results from mirror planes that are parallel to the 6-fold axis.

Dipyramids

Dipyramids are closed forms consisting of 6, 8, 12, 16, or 24 faces. Dipyramids are pyramids that are reflected across a mirror plane. Thus, they occur in crystal classes that have a mirror plane perpendicular to a rotation or rotoinversion axis.

- Trigonal dipyramid: 6-faced form with faces related by a 3-fold axis with a perpendicular mirror plane. In this drawing, all six faces belong to the trigonal-dipyramid.

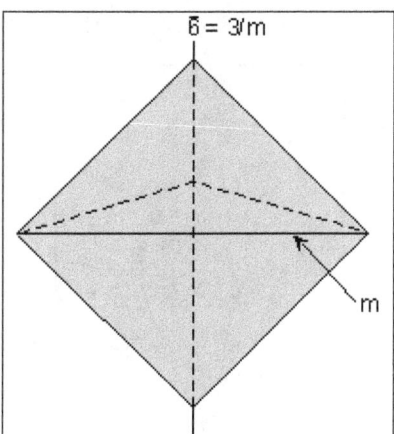

- Ditrigonal-dipyramid: 12-faced form with faces related by a 3-fold axis with a perpendicular mirror plane. If viewed from above, the crystal will not have a hexagonal shape, rather it would appear similar to the horizontal cross-section of the ditrigonal prism, discussed above.

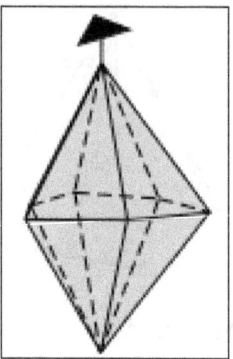

- Rhombic dipyramid: 8-faced form with faces related by a combinations of 2-fold axes and mirror planes. The drawing to the right shows 2 rhombic dipyramids. One has the form symbol {111} and consists of the four larger faces shown plus four equivalent faces on the back of the model. The other one has the form symbol {113} and consists of the 4 smaller faces shown plus the four on the back.

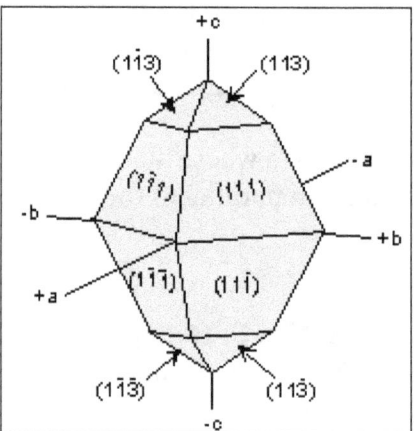

- Tetragonal dipyramid: 8-faced form with faces related by a 4-fold axis with a perpendicular mirror plane. The drawing shows the 8-faced tetragonal dipyramid. Also shown is the 4-faced tetragonal prism, and the 2-faced top/bottom pinacoid.

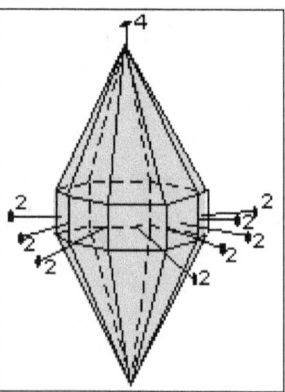

- Ditetragonal dipyramid: 16-faced form with faces related by a 4-fold axis with a perpendicular mirror plane. The ditetragonal dipyramid is shown here. Note the vertical faces belong to a ditetragonal prism.

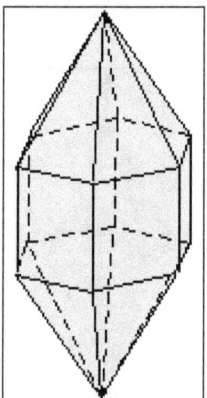

- Hexagonal dipyramid: 12-faced form with faces related by a 6-fold axis with a perpendicular mirror plane. The vertical faces in this model make up a hexagonal prism.

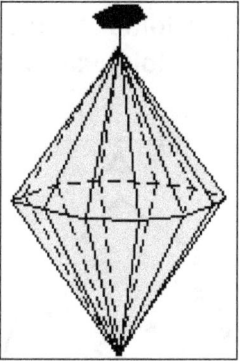

- Dihexagonal dipyramid: 24-faced form with faces related by a 6-fold axis with a perpendicular mirror plane.

Trapezohedrons

Trapezohedron are closed 6, 8, or 12 faced forms, with 3, 4, or 6 upper faces offset from 3, 4, or 6 lower faces. The trapezohedron results from 3-, 4-, or 6-fold axes combined with a perpendicular 2-fold axis. An example of a tetragonal trapezohedron is shown in the drawing to the right.

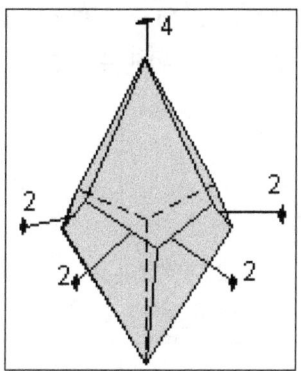

Scalenohedrons

A scalenohedron is a closed form with 8 or 12 faces. In ideally developed faces each of the faces is a scalene triangle. In the model, note the presence of the 3-fold rotoinversion axis perpendicular to the 3 2-fold axes.

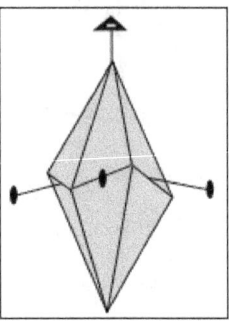

Rhombohedrons

A rhombohedron is 6-faced closed form wherein 3 faces on top are offset by 3 identical upside down faces on the bottom, as a result of a 3-fold rotoinversion axis. Rhombohedrons can also result from a 3-fold axis with perpendicular 2-fold axes. Rhombohedrons only occur in the crystal classes 2/m, 32, and 3.

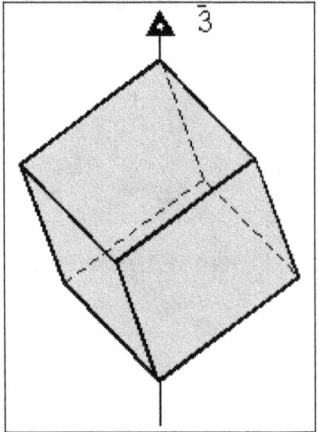

Disphenoids

A disphenoid is a closed form consisting of 4 faces. These are only present in the orthorhombic system (class 222) and the tetragonal system (class $\bar{4}$).

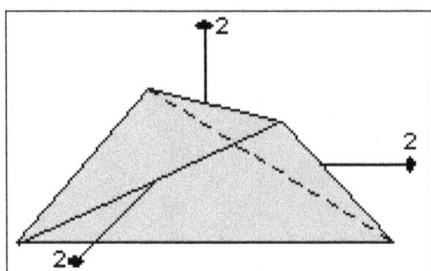

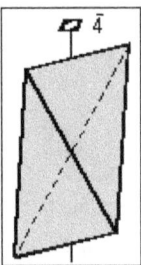

The rest of the forms all occur in the isometric system, and thus have either four 3-fold axes or four axes.

Hexahedron

A hexahedron is the same as a cube. 4-fold axes are perpendicular to the face of the cube, and four $\bar{3}$ axes run through the corners of the cube. Note that the form symbol for a hexahedron is {100}, and it consists of the following 6 faces:

$$(100), (010), (001), (\bar{1}00), (0\bar{1}0), \text{ and } (00\bar{1}).$$

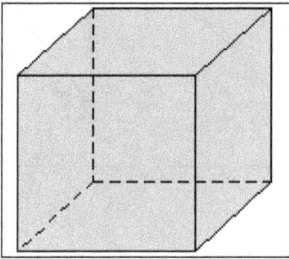

Octahedron

An octahedron is an 8 faced form that results form three 4-fold axes with perpendicular mirror planes. The octahedron has the form symbol {111}and consists of the following 8 faces:

$$(111), (\bar{1}\bar{1}\bar{1}), (1\bar{1}1), (1\bar{1}\bar{1}), (\bar{1}\bar{1}1), (\bar{1}1\bar{1}), (11\bar{1}), \text{ and } (\bar{1}11).$$

Note that four 3-fold axes are present that are perpendicular to the triangular faces of the octahedron (these 3-fold axes are not shown in the drawing).

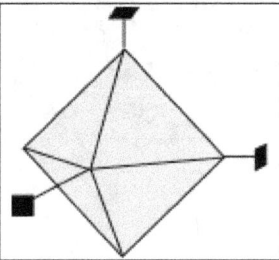

Dodecahedron

A dodecahedron is a closed 12-faced form. Dodecahedrons can be formed by cutting off the edges of a cube. The form symbol for a dodecahedron is {110}.

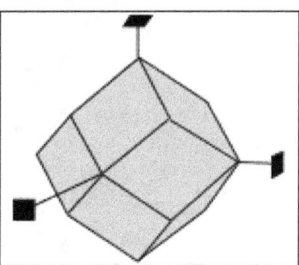

Tetrahexahedron

The tetrahexahedron is a 24-faced form with a general form symbol of {ohl}. This means that all faces are parallel to one of the a axes, and intersect the other 2 axes at different lengths.

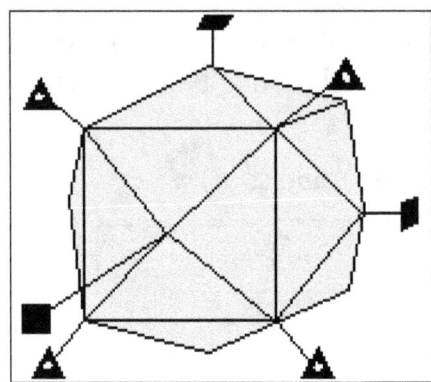

Trapezohedron

An isometric trapezohedron is a 12-faced closed form with the general form symbol {hhl}. This means that all faces intersect two of the a axes at equal length and intersect the third a axis at a different length.

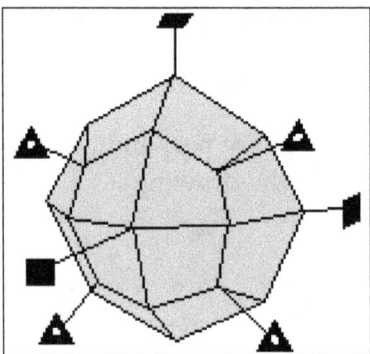

Tetrahedron

The tetrahedron occurs in the class 3m and has the form symbol {111}(the form shown in the drawing) or {111} (2 different forms are possible). It is a four faced form that results form three axes and four 3-fold axes (not shown in the drawing).

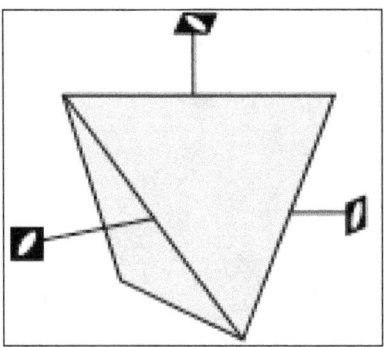

Gyroid

A gyroid is a form in the class 432 (note no mirror planes).

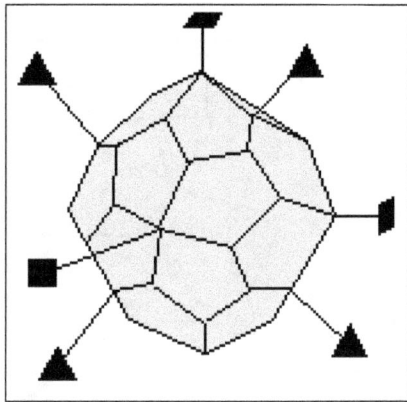

Pyritohedron

The pyritohedron is a 12-faced form that occurs in the crystal class 2/m. Note that there are no 4-fold axes in this class. The possible forms are {h0l} or {0kl} and each of the faces that make up the form have 5 sides.

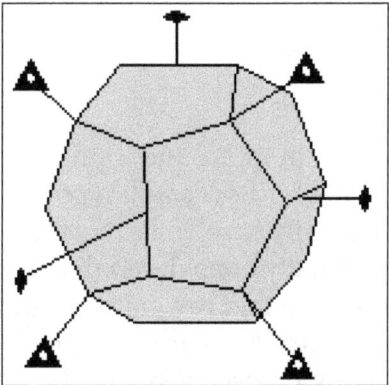

Diploid

The diploid is the general form {hkl} for the diploidal class (2/m). Again there are no 4-fold axes.

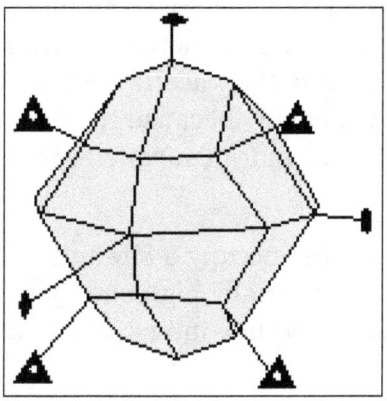

Tetartoid

Tetartoids are general forms in the tetartoidal class (23) which only has 3-fold axes and 2-fold axes with no mirror planes.

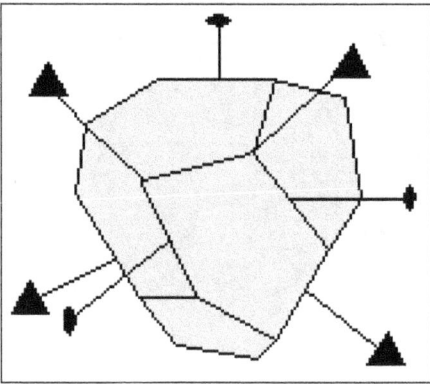

Crystallographic Defect

Crystallographic defects are interruptions of regular patterns in crystalline solids. They are common because positions of atoms or molecules at repeating fixed distances determined by the unit cell parameters in crystals, which exhibit a periodic crystal structure, are usually imperfect.

Point Defects

Point defects are defects that occur only at or around a single lattice point. They are not extended in space in any dimension. Strict limits for how small a point defect is are generally not defined explicitly. However, these defects typically involve at most a few extra or missing atoms. Larger defects in an ordered structure are usually considered dislocation loops. For historical reasons, many point defects, especially in ionic crystals, are called centers: for example a vacancy in many ionic solids is called a luminescence center, a color center, or F-center. These dislocations permit ionic transport through crystals leading to electrochemical reactions. These are frequently specified using Kröger–Vink notation.

- Vacancy defects are lattice sites which would be occupied in a perfect crystal, but are vacant. If a neighboring atom moves to occupy the vacant site, the vacancy moves in the opposite direction to the site which used to be occupied by the moving atom. The stability of the surrounding crystal structure guarantees that the neighboring atoms will not simply collapse around the vacancy. In some materials, neighboring atoms actually move away from a vacancy, because they experience attraction from atoms in the surroundings. A vacancy (or pair of vacancies in an ionic solid) is sometimes called a Schottky defect.

- Interstitial defects are atoms that occupy a site in the crystal structure at which there is usually not an atom. They are generally high energy configurations. Small atoms (mostly impurities) in some crystals can occupy interstices without high energy, such as hydrogen in palladium.

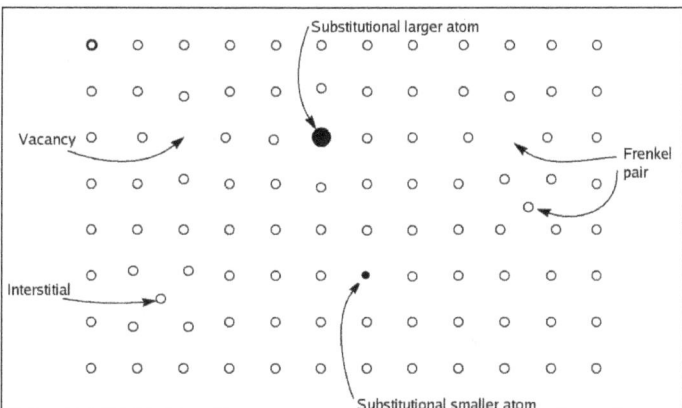

Schematic illustration of some simple point defect types in a monatomic solid.

- A nearby pair of a vacancy and an interstitial is often called a Frenkel defect or Frenkel pair. This is caused when an ion moves into an interstitial site and creates a vacancy.

- Due to fundamental limitations of material purification methods, materials are never 100% pure, which by definition induces defects in crystal structure. In the case of an impurity, the atom is often incorporated at a regular atomic site in the crystal structure. This is neither a vacant site nor is the atom on an interstitial site and it is called a substitutional defect. The atom is not supposed to be anywhere in the crystal, and is thus an impurity. In some cases where the radius of the substitutional atom (ion) is substantially smaller than that of the atom (ion) it is replacing, its equilibrium position can be shifted away from the lattice site. These types of substitutional defects are often referred to as off-center ions. There are two different types of substitutional defects: Isovalent substitution and aliovalent substitution. Isovalent substitution is where the ion that is substituting the original ion is of the same oxidation state as the ion it is replacing. Aliovalent substitution is where the ion that is substituting the original ion is of a different oxidation state than the ion it is replacing. Aliovalent substitutions change the overall charge within the ionic compound, but the ionic compound must be neutral. Therefore, a charge compensation mechanism is required. Hence either one of the metals is partially or fully oxidised or reduced, or ion vacancies are created.

- Antisite defects occur in an ordered alloy or compound when atoms of different type exchange positions. For example, some alloys have a regular structure in which every other atom is a different species; for illustration assume that type A atoms sit on the corners of a cubic lattice, and type B atoms sit in the center of the cubes. If one cube has an A atom at its center, the atom is on a site usually occupied by a B atom, and is thus an antisite defect. This is neither a vacancy nor an interstitial, nor an impurity.

- Topological defects are regions in a crystal where the normal chemical bonding environment is topologically different from the surroundings. For instance, in a perfect sheet of graphite (graphene) all atoms are in rings containing six atoms. If the sheet contains regions where the number of atoms in a ring is different from six, while the total number of atoms remains the same, a topological defect has formed. An example is the Stone Wales defect in nanotubes, which consists of two adjacent 5-membered and two 7-membered atom rings.

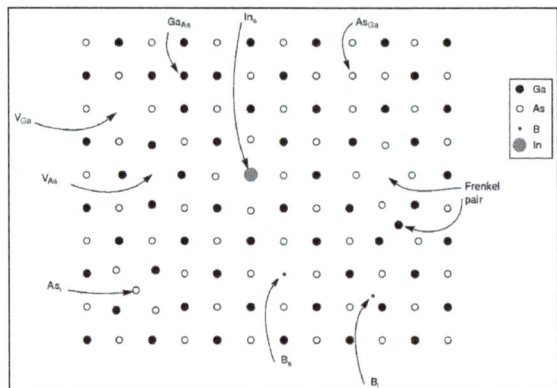

Schematic illustration of defects in a compound solid, using GaAs as an example.

- Also amorphous solids may contain defects. These are naturally somewhat hard to define, but sometimes their nature can be quite easily understood. For instance, in ideally bonded amorphous silica all Si atoms have 4 bonds to O atoms and all O atoms have 2 bonds to Si atom. Thus e.g. an O atom with only one Si bond (a dangling bond) can be considered a defect in silica. Moreover, defects can also be defined in amorphous solids based on empty or densely packed local atomic neighbourhoods, and the properties of such 'defects' can be shown to be similar to normal vacancies and interstitials in crystals.

- Complexes can form between different kinds of point defects. For example, if a vacancy encounters an impurity, the two may bind together if the impurity is too large for the lattice. Interstitials can form 'split interstitial' or 'dumbbell' structures where two atoms effectively share an atomic site, resulting in neither atom actually occupying the site.

Line defects

Line defects can be described by gauge theories.

Dislocations are linear defects, around which the atoms of the crystal lattice are misaligned. There are two basic types of dislocations, the edge dislocation and the screwdislocation. "Mixed" dislocations, combining aspects of both types, are also common.

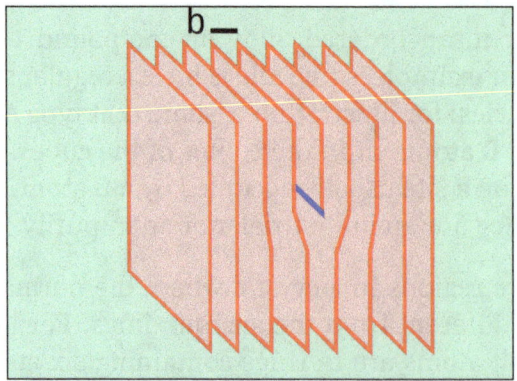

An edge dislocation is shown. The dislocation line is presented in blue, the Burgers vector b in black.

Edge dislocations are caused by the termination of a plane of atoms in the middle of a crystal. In such a case, the adjacent planes are not straight, but instead bend around the edge of the terminating plane so that the crystal structure is perfectly ordered on either side. The analogy with a stack of paper is apt: if a half a piece of paper is inserted in a stack of paper, the defect in the stack is only noticeable at the edge of the half sheet.

The screw dislocation is more difficult to visualise, but basically comprises a structure in which a helical path is traced around the linear defect (dislocation line) by the atomic planes of atoms in the crystal lattice.

The presence of dislocation results in lattice strain (distortion). The direction and magnitude of such distortion is expressed in terms of a Burgers vector (b). For an edge type, b is perpendicular to the dislocation line, whereas in the cases of the screw type it is parallel. In metallic materials, b is aligned with close-packed crystallographic directions and its magnitude is equivalent to one interatomic spacing.

Dislocations can move if the atoms from one of the surrounding planes break their bonds and re-bond with the atoms at the terminating edge.

It is the presence of dislocations and their ability to readily move (and interact) under the influence of stresses induced by external loads that leads to the characteristic malleabilityof metallic materials.

Dislocations can be observed using transmission electron microscopy, field ion microscopy and atom probe techniques. Deep-level transient spectroscopy has been used for studying the electrical activity of dislocations in semiconductors, mainly silicon.

Disclinations are line defects corresponding to "adding" or "subtracting" an angle around a line. Basically, this means that if you track the crystal orientation around the line defect, you get a rotation. Usually, they were thought to play a role only in liquid crystals, but recent developments suggest that they might have a role also in solid materials, e.g. leading to the self-healing of cracks.

Planar Defects

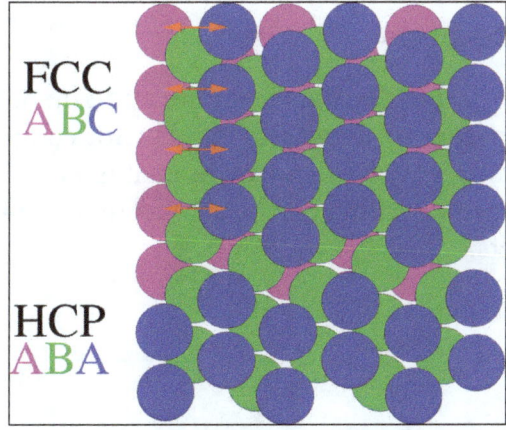

Origin of stacking faults: Different stacking sequences of close-packed crystals.

- Grain boundaries occur where the crystallographic direction of the lattice abruptly changes. This usually occurs when two crystals begin growing separately and then meet.

- Antiphase boundaries occur in ordered alloys: in this case, the crystallographic direction remains the same, but each side of the boundary has an opposite phase: For example, if the ordering is usually ABABABAB (hexagonal close-packed crystal), an antiphase boundary takes the form of ABABBABA.

- Stacking faults occur in a number of crystal structures, but the common example is in close-packed structures. They are formed by a local deviation of the stacking sequence of layers in a crystal. An example would be the ABABCABAB stacking sequence.

- A twin boundary is a defect that introduces a plane of mirror symmetry in the ordering of a crystal. For example, in cubic close-packed crystals, the stacking sequence of a twin boundary would be ABCABCBACBA.

- On planes of single crystals, steps between atomically flat terraces can also be regarded as planar defects. It has been shown that such defects and their geometry have significant influence on the adsorption of organic molecules.

Bulk Defects

- Three-Dimensional macroscopic or bulk defects, such as pores, cracks, or inclusions.

- Voids — Small regions where there are no atoms, and which can be thought of as clusters of vacancies.

- Impurities can cluster together to form small regions of a different phase. These are often called precipitates.

Mathematical Classification Methods

A successful mathematical classification method for physical lattice defects, which works not only with the theory of dislocations and other defects in crystals but also, e.g., for disclinations in liquid crystals and for excitations in superfluid ^{3}He, is the topological homotopy theory.

Computer Simulation Methods

Density functional theory, classical molecular dynamics and kinetic Monte Carlo simulations are widely used to study the properties of defects in solids with computer simulations. Simulating jamming of hard spheres of different sizes and/or in containers with non-commeasurable sizes using the Lubachevsky–Stillinger algorithm can be an effective technique for demonstrating some types of crystallographic defects.

Lattice Plane

In crystallography, a lattice plane of a given Bravais lattice is a plane (or family of parallel planes) whose intersections with the lattice (or any crystalline structure of that lattice) are periodic (i.e. are described by 2d Bravais lattices) and intersect the Bravais lattice; equivalently, a lattice plane is any plane containing at least three noncollinear Bravais lattice points. All lattice planes can be described by a set of integer Miller indices, and vice versa (all integer Miller indices define lattice planes).

Conversely, planes that are not lattice planes have aperiodic intersections with the lattice called quasicrystals; this is known as a "cut-and-project" construction of a quasicrystal (and is typically also generalized to higher dimensions).

Steno's Law

Steno's law states that the angles between two corresponding faces on the crystals of any solid chemical or mineral species are constant and are characteristic of the species; this angle is measured between lines drawn perpendicular to each face. The law, also called the law of constancy of interfacial angles, holds for any two crystals, regardless of size, locality of occurrence, or whether they are natural or man-made.

The relationship was discovered in 1669 by the Danish geologist Nicolaus Steno, who noted that, although quartz crystals differ in appearance from one to another, the angles between corresponding faces are always the same. In 1772 a French mineralogist, Jean-Baptiste L. Romé de l'Isle, confirmed Steno's findings and further noted that the angles are characteristic of the substance. A French crystallographer, René-Just Haüy, usually considered the father of crystallography, showed in 1774 that the known interfacial angles could be accounted for if the crystal were made up of minute building blocks that correspond to the present-day unit cells.

Miller Index

Miller indices form a notation system in crystallography for planes in crystal (Bravais) lattices.

In particular, a family of lattice planes is determined by three integers h, k, and ℓ, the *Miller indices*. They are written (hkℓ), and denote the family of planes orthogonal to $h b_1 + k b_2 + \ell b_3$, where b_i are the basis of the reciprocal lattice vectors. (Note that the plane is not always orthogonal to the linear combination of direct lattice vectors $h a_1 + k a_2 + \ell a_3$ because the reciprocal lattice vectors need not be mutually orthogonal.) By convention, negative integers are written with a bar, as in 3 for −3. The integers are usually written in lowest terms, i.e. their greatest common divisor should be 1.

There are also several related notations:

- The notation {hkℓ} denotes the set of all planes that are equivalent to (hkℓ) by the symmetry of the lattice.

In the context of crystal *directions* (not planes), the corresponding notations are:

- [hkℓ], with square instead of round brackets, denotes a direction in the basis of the *direct* lattice vectors instead of the reciprocal lattice; and

- Similarly, the notation <hkℓ> denotes the set of all directions that are equivalent to [hkℓ] by symmetry.

Miller indices were introduced in 1839 by the British mineralogist William Hallowes Miller, although an almost identical system (Weiss parameters) had already been used by German mineralogist Christian Samuel Weiss since 1817. The method was also historically known as the Millerian system, and the indices as Millerian, although this is now rare.

The Miller indices are defined with respect to any choice of unit cell and not only with respect to primitive basis vectors, as is sometimes stated.

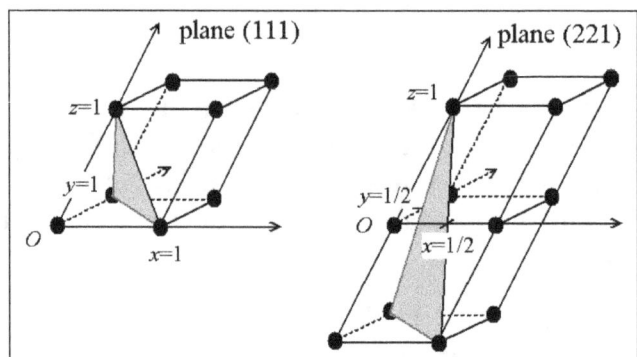

Examples of determining indices for a plane using intercepts with axes; left (111), right (221)

There are two equivalent ways to define the meaning of the Miller indices: via a point in the reciprocal lattice, or as the inverse intercepts along the lattice vectors. Both definitions are given below. In either case, one needs to choose the three lattice vectors a_1, a_2, and a_3 that define the unit cell (note that the conventional unit cell may be larger than the primitive cell of the Bravais lattice, as the examples below illustrate). Given these, the three primitive reciprocal lattice vectors are also determined (denoted b_1, b_2, and b_3).

Then, given the three Miller indices h, k, ℓ, (hkℓ) denotes planes orthogonal to the reciprocal lattice vector:

$$g_{hk\ell} = h\mathrm{b}_1 + k\mathrm{b}_2 + \ell\mathrm{b}_3.$$

That is, (hkℓ) simply indicates a normal to the planes in the basis of the primitive reciprocal lattice vectors. Because the coordinates are integers, this normal is itself always a reciprocal lattice vector. The requirement of lowest terms means that it is the shortest reciprocal lattice vector in the given direction.

Equivalently, (hkℓ) denotes a plane that intercepts the three points a_1/h, a_2/k, and a_3/ℓ, or some multiple thereof. That is, the Miller indices are proportional to the *inverses* of the intercepts of the plane, in the basis of the lattice vectors. If one of the indices is zero, it means that the planes do not intersect that axis (the intercept is "at infinity").

Considering only (hkℓ) planes intersecting one or more lattice points (the lattice planes), the perpendicular distance d between adjacent lattice planes is related to the (shortest) reciprocal lattice vector orthogonal to the planes by the formula: $d = 2\pi/|g_{hk\ell}|$..

The related notation [hkℓ] denotes the direction:

$$h\mathrm{a}_1 + k\mathrm{a}_2 + \ell\mathrm{a}_3.$$

That is, it uses the direct lattice basis instead of the reciprocal lattice. Note that [hkℓ] is not generally normal to the (hkℓ) planes, except in a cubic lattice as described below.

Case of Cubic Structures

For the special case of simple cubic crystals, the lattice vectors are orthogonal and of equal length (usually denoted a), as are those of the reciprocal lattice. Thus, in this common case, the Miller indices (hkℓ) and [hkℓ] both simply denote normals/directions in Cartesian coordinates.

For cubic crystals with lattice constant a, the spacing d between adjacent (hkℓ) lattice planes is (from above)

$$d_{hk\ell} = \frac{a}{\sqrt{h^2 + k^2 + \ell^2}}.$$

Because of the symmetry of cubic crystals, it is possible to change the place and sign of the integers and have equivalent directions and planes:

- Indices in *angle brackets* such as ⟨100⟩ denote a *family* of directions which are equivalent due to symmetry operations, such as [100], [010], or the negative of any of those directions.

- Indices in *curly brackets* or *braces* such as {100} denote a family of plane normals which are equivalent due to symmetry operations, much the way angle brackets denote a family of directions.

For face-centered cubic and body-centered cubic lattices, the primitive lattice vectors are not orthogonal. However, in these cases the Miller indices are conventionally defined relative to the lattice vectors of the cubic supercell and hence are again simply the Cartesian directions.

Case of Hexagonal and Rhombohedral Structures

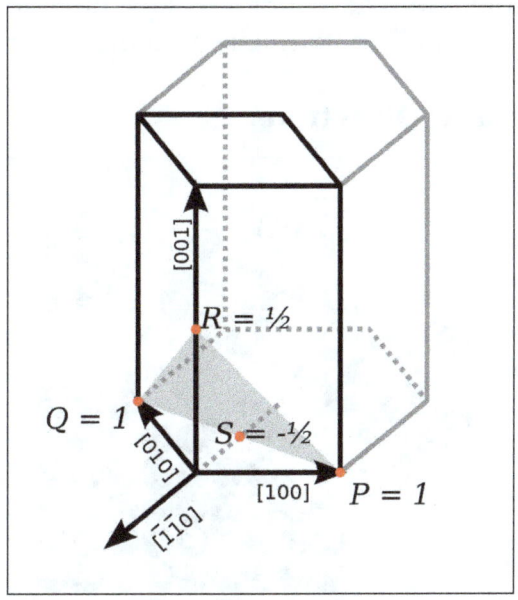

Miller-Bravais indices.

With hexagonal and rhombohedral lattice systems, it is possible to use the Bravais-Miller system, which uses four indices $(h\,k\,i\,\ell)$ that obey the constraint.

$$h + k + i = 0.$$

Here h, k and ℓ are identical to the corresponding Miller indices, and i is a redundant index.

This four-index scheme for labeling planes in a hexagonal lattice makes permutation symmetries apparent. For example, the similarity between (110) ≡ (1120) and (120) ≡ (1210) is more obvious when the redundant index is shown.

The (001) plane has a 3-fold symmetry: it remains unchanged by a rotation of 1/3 ($2\pi/3$ rad, 120°). The [100], and the [$\bar{1}\bar{1}$0] directions are really similar. If S is the intercept of the plane with the [$\bar{1}\bar{1}$0] axis, then:

$$i = 1/S.$$

There are also *ad hoc* schemes (e.g. in the transmission electron microscopy literature) for indexing hexagonal *lattice vectors* (rather than reciprocal lattice vectors or planes) with four indices. However they don't operate by similarly adding a redundant index to the regular three-index set.

For example, the reciprocal lattice vector (hkℓ) as suggested above can be written in terms of reciprocal lattice vectors as $h\,b_1 + k\,b_2 + \ell\,b_3$. For hexagonal crystals this may be expressed in terms of direct-lattice basis-vectors a_1, a_2 and a_3 as:

$$h\,b_1 + k b_2 + \ell b_3 = \frac{2}{3a^2}(2h+k)\,a_1 + \frac{2}{3a^2}(h+2k)\,a_2 + \frac{1}{c^2}(\ell)\,a_3\,.$$

Hence zone indices of the direction perpendicular to plane (hkℓ) are, in suitably normalized triplet form, simply $[2h+k, h+2k, \ell(3/2)(a/c)^2]$. When *four indices* are used for the zone normal to plane (hkℓ), however, the literature often uses $[h, k, -h-k, \ell(3/2)(a/c)^2]$ instead. Thus as you can see, four-index zone indices in square or angle brackets sometimes mix a single direct-lattice index on the right with reciprocal-lattice indices (normally in round or curly brackets) on the left.

Crystallographic Planes and Directions

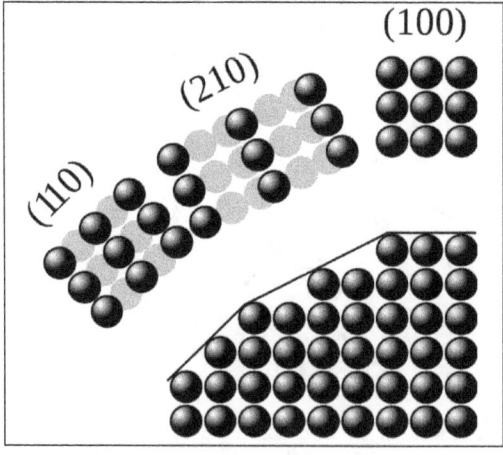

Dense crystallographic planes

Crystallographic directions are fictitious lines linking nodes (atoms, ions or molecules) of a crystal. Similarly, crystallographic planes are fictitious *planes* linking nodes. Some directions and planes have a higher density of nodes; these dense planes have an influence on the behavior of the crystal:

- Optical properties: In condensed matter, light "jumps" from one atom to the other with the Rayleigh scattering; the velocity of light thus varies according to the directions, whether the atoms are close or far; this gives the birefringence.

- Adsorption and reactivity: Adsorption and chemical reactions can occur at atoms or molecules on crystal surfaces, these phenomena are thus sensitive to the density of nodes;

- Surface tension: The condensation of a material means that the atoms, ions or molecules are more stable if they are surrounded by other similar species; the surface tension of an interface thus varies according to the density on the surface :

  ◦ Pores and crystallites tend to have straight grain boundaries following dense planes.

  ◦ Cleavage.

- Dislocations (plastic deformation) :

  ◦ The dislocation core tends to spread on dense planes (the elastic perturbation is "diluted"); this reduces the friction (Peierls–Nabarro force), the sliding occurs more frequently on dense planes.

  ◦ The perturbation carried by the dislocation (Burgers vector) is along a dense direction: the shift of one node in a dense direction is a lesser distortion.

  ◦ The dislocation line tends to follow a dense direction, the dislocation line is often a straight line, a dislocation loop is often a polygon.

For all these reasons, it is important to determine the planes and thus to have a notation system.

Integer vs. Irrational Miller Indices: Lattice Planes and Quasicrystals

Ordinarily, Miller indices are always integers by definition, and this constraint is physically significant. To understand this, suppose that we allow a plane (abc) where the Miller "indices" a, b and c (defined as above) are not necessarily integers.

If a, b and c have rational ratios, then the same family of planes can be written in terms of integer indices (hkℓ) by scaling a, b and c appropriately: divide by the largest of the three numbers, and then multiply by the least common denominator. Thus, integer Miller indices implicitly include indices with all rational ratios. The reason why planes where the components (in the reciprocal-lattice basis) have rational ratios are of special interest is that these are the lattice planes: they are the only planes whose intersections with the crystal are 2d-periodic.

For a plane (abc) where a, b and c have irrational ratios, on the other hand, the intersection of the plane with the crystal is *not* periodic. It forms an aperiodic pattern known as a quasicrystal. This construction corresponds precisely to the standard "cut-and-project" method of defining a quasicrystal, using a plane with irrational-ratio Miller indices. (Although many quasicrystals, such as the

Penrose tiling, are formed by "cuts" of periodic lattices in more than three dimensions, involving the intersection of more than one such hyperplane.)

Law of Rational Indices

The law of rational indices states that the intercepts, OP, OQ, OR, of the natural faces of a crystal form with the unit-cell axes a, b, c are inversely proportional to prime integers, h, k, l. They are called the Miller indices of the face. They are usually small because the corresponding lattice planes are among the densest and have therefore a high interplanar spacing and low indices.

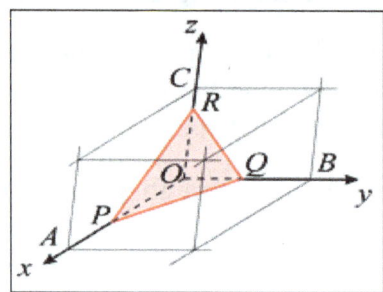

The law of rational indices was deduced by Haüy from the observation of the stacking laws required to build the natural faces of crystals by piling up elementary blocks, for instance cubes to construct the {110} faces of the rhomb-dodecahedron observed in garnets or the ½{210} faces of the pentagon-dodecahedron observed in pyrite, or rhombohedrons to construct the {21.1} (referred to an hexagonal lattice, {2$\bar{1}$0}, referred to a rhombohedral lattice) scalenohedron of calcite.

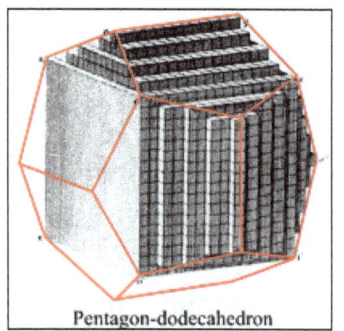
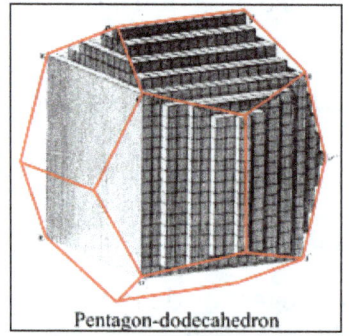
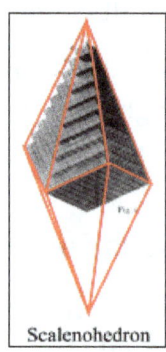

Pentagon-dodecahedron

Pentagon-dodecahedron

Scalenohedron

Crystallographic Restriction Theorem

The crystallographic restriction theorem in its basic form was based on the observation that the rotational symmetries of a crystal are usually limited to 2-fold, 3-fold, 4-fold, and 6-fold. However, quasicrystals can occur with other diffraction pattern symmetries, such as 5-fold; these were not discovered until 1982 by Dan Shechtman.

Crystals are modeled as discrete lattices, generated by a list of independent finite translations. Because discreteness requires that the spacings between lattice points have a lower bound, the group of rotational symmetries of the lattice at any point must be a finite group (alternatively, the point is the only system allowing for infinite rotational symmetry). The strength of the theorem is that not all finite groups are compatible with a discrete lattice; in any dimension, we will have only a finite number of compatible groups.

Dimensions 2 and 3

The special cases of 2D (wallpaper groups) and 3D (space groups) are most heavily used in applications, and they can be treated together.

Lattice Proof

A rotation symmetry in dimension 2 or 3 must move a lattice point to a succession of other lattice points in the same plane, generating a regular polygon of coplanar lattice points.

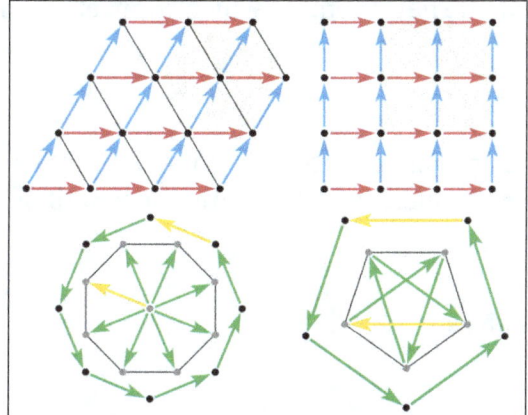

Lattices restrict polygons Compatible: 6-fold (3-fold), 4-fold (2-fold)
Incompatible: 8-fold, 5-fold.

Now consider an 8-fold rotation, and the displacement vectors between adjacent points of the polygon. If a displacement exists between any two lattice points, then that same displacement is repeated everywhere in the lattice. So collect all the edge displacements to begin at a single lattice point. The edge vectors become radial vectors, and their 8-fold symmetry implies a regular octagon of lattice points around the collection point. But this is *impossible*, because the new octagon is about 80% as large as the original. The significance of the shrinking is that it is unlimited. The same construction can be repeated with the new octagon, and again and again until the distance between lattice points is as small as we like; thus no *discrete* lattice can have 8-fold symmetry. The same argument applies to any k-fold rotation, for k greater than 6.

A shrinking argument also eliminates 5-fold symmetry. Consider a regular pentagon of lattice points. If it exists, then we can take every *other* edge displacement and (head-to-tail) assemble a 5-point star, with the last edge returning to the starting point. The vertices of such a star are again vertices of a regular pentagon with 5-fold symmetry, but about 60% smaller than the original.

Thus the theorem is proved.

The existence of quasicrystals and Penrose tilings shows that the assumption of a linear translation is necessary. Penrose tilings may have 5-fold rotational symmetry and a discrete lattice, and any local neighborhood of the tiling is repeated infinitely many times, but there is no linear translation for the tiling as a whole. And without the discrete lattice assumption, the above construction not only fails to reach a contradiction, but produces a (non-discrete) counterexample. Thus 5-fold rotational symmetry cannot be eliminated by an argument missing either of those assumptions. A Penrose tiling of the whole (infinite) plane can only have exact 5-fold rotational symmetry (of the whole tiling) about a single point, however, whereas the 4-fold and 6-fold lattices have infinitely many centres of rotational symmetry.

Trigonometry Proof

Consider two lattice points A and B separated by a translation vector r. Consider an angle α such that a rotation of angle α about any lattice point is a symmetry of the lattice. Rotating about point B by α maps point A to a new point A'. Similarly, rotating about point A by α maps B to a point B'. Since both rotations mentioned are symmetry operations, A' and B' must both be lattice points. Due to periodicity of the crystal, the new vector r' which connects them must be equal to an integer multiple of r:

$$r' = mr$$

Here m is an integer. The four translation vectors, three of length $r = |r|$ and one, connecting A' and B', of length $'\ |r'|$, form a trapezium. Therefore, the length of r' is also given by:

$$r' = 2r \cos\alpha - r.$$

Combining the two equations gives:

$$\cos\alpha = \frac{m+1}{2} = \frac{M}{2}$$

Here $M = m+1$ is also an integer. Bearing in mind that $|\cos\alpha| \leq 1$ we have allowed integers $M \in \{-2, -1, 0, 1, 2\}$. Solving for possible values of $M \in \{-2, -1, 0, 1, 2\}$ reveals that the only values in the 0° to 180° range are 0°, 60°, 90°, 120°, and 180°. In radians, the only allowed rotations consistent with lattice periodicity are given by $2\pi/n$, where n = 1, 2, 3, 4, 6. This corresponds to 1-, 2-, 3-, 4-, and 6-fold symmetry, respectively, and therefore excludes the possibility of 5-fold or greater than 6-fold symmetry.

Short Trigonometry Proof

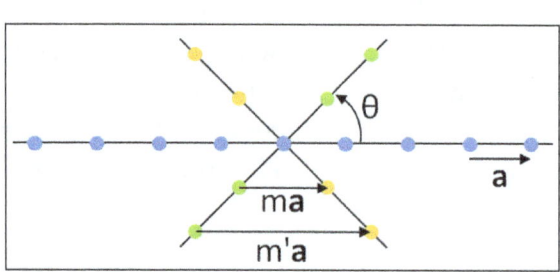 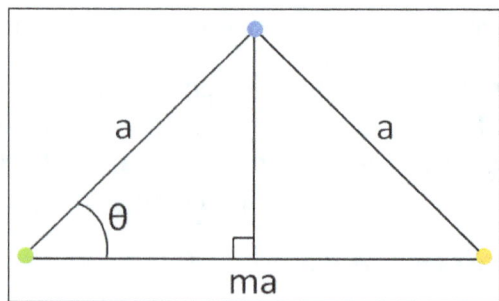

Consider a line of atoms A-O-B, separated by distance a. Rotate the entire row by $\theta = +2\pi/n$ and $\theta = -2\pi/n$, with point O kept fixed. After the rotation by $+2\pi/n$, A is moved to the lattice point C and after the rotation by -2π/n, B is moved to the lattice point D. Due to the assumed periodicity of the lattice, the two lattice points C and D will be also in a line directly below the initial row; moreover C and D will be separated by $r = ma$, with m an integer. But by the geometry, the separation between these points is:

$$2a \cos \theta = 2a \cos \frac{2\pi}{n}.$$

Equating the two relations gives:

$$2 \cos \frac{2\pi}{n} = m$$

This is satisfied by only n = 1, 2, 3, 4, 6.

Matrix Proof

For an alternative proof, consider matrix properties. The sum of the diagonal elements of a matrix is called the trace of the matrix. In 2D and 3D every rotation is a planar rotation, and the trace is a function of the angle alone. For a 2D rotation, the trace is $2 \cos \theta$; for a 3D rotation, $1 + 2 \cos \theta$.

Examples:

- Consider a 60° (6-fold) rotation matrix with respect to an orthonormal basis in 2D.

$$\begin{bmatrix} 1/2 & -\sqrt{3}/2 \\ \sqrt{3}/2 & 1/2 \end{bmatrix}$$

   The trace is precisely 1, an integer.

- Consider a 45° (8-fold) rotation matrix.

$$\begin{bmatrix} 1/\sqrt{2} & -1/\sqrt{2} \\ 1/\sqrt{2} & 1/\sqrt{2} \end{bmatrix}$$

   The trace is $2/\sqrt{2}$, not an integer.

Selecting a basis formed from vectors that spans the lattice, neither orthogonality nor unit length is guaranteed, only linear independence. However the trace of the rotation matrix is the same with respect to any basis. The trace is a similarity invariant under linear transformations. In the lattice basis, the rotation operation must map every lattice point into an integer number of lattice vectors, so the entries of the rotation matrix in the lattice basis — and hence the trace — are necessarily integers. Similar as in other proofs, this implies that the only allowed rotational symmetries correspond to 1, 2, 3, 4 or 6-fold invariance. For example, wallpapers and crystals cannot be rotated by 45° and remain invariant, the only possible angles are: 360°, 180°, 120°, 90° or 60°.

Example:

Consider a 60° (360°/6) rotation matrix with respect to the oblique lattice basis for a tiling by equilateral triangles.

$$\begin{bmatrix} 0 & -1 \\ 1 & 1 \end{bmatrix}$$

The trace is still 1. The determinant (always +1 for a rotation) is also preserved.

The general crystallographic restriction on rotations does *not* guarantee that a rotation will be compatible with a specific lattice. For example, a 60° rotation will not work with a square lattice; nor will a 90° rotation work with a rectangular lattice.

Higher Dimensions

When the dimension of the lattice rises to four or more, rotations need no longer be planar; the 2D proof is inadequate. However, restrictions still apply, though more symmetries are permissible. For example, the hypercubic lattice has an eightfold rotational symmetry, corresponding to an eightfold rotational symmetry of the hypercube. This is of interest, not just for mathematics, but for the physics of quasicrystals under the cut-and-project theory. In this view, a 3D quasicrystal with 8-fold rotation symmetry might be described as the projection of a slab cut from a 4D lattice.

The following 4D rotation matrix is the aforementioned eightfold symmetry of the hypercube (and the cross-polytope):

$$A = \begin{bmatrix} 0 & 0 & 0 & -1 \\ 1 & 0 & 0 & 0 \\ 0 & -1 & 0 & 0 \\ 0 & 0 & -1 & 0 \end{bmatrix}.$$

Transforming this matrix to the new coordinates given by

$$B = \begin{bmatrix} -1/2 & 0 & -1/2 & \sqrt{2}/2 \\ 1/2 & \sqrt{2}/2 & -1/2 & 0 \\ -1/2 & 0 & -1/2 & -\sqrt{2}/2 \\ -1/2 & \sqrt{2}/2 & 1/2 & 0 \end{bmatrix}$$ will produce:

$$BAB^{-1} = \begin{bmatrix} \sqrt{2}/2 & \sqrt{2}/2 & 0 & 0 \\ -\sqrt{2}/2 & \sqrt{2}/2 & 0 & 0 \\ 0 & 0 & -\sqrt{2}/2 & \sqrt{2}/2 \\ 0 & 0 & -\sqrt{2}/2 & -\sqrt{2}/2 \end{bmatrix}.$$

This third matrix then corresponds to a rotation both by 45° (in the first two dimensions) and by 135° (in the last two). Projecting a slab of hypercubes along the first two dimensions of the new coordinates produces an Ammann–Beenker tiling (another such tiling is produced by projecting along the last two dimensions), which therefore also has 8-fold rotational symmetry on average.

The A4 lattice and F4 lattice have order 10 and order 12 rotational symmetries, respectively.

To state the restriction for all dimensions, it is convenient to shift attention away from rotations alone and concentrate on the integer matrices. We say that a matrix A has order k when its k-th power (but no lower), A^k, equals the identity. Thus a 6-fold rotation matrix in the equilateral triangle basis is an integer matrix with order 6. Let Ord_N denote the set of integers that can be the order of an $N \times N$ integer matrix. For example, $\mathrm{Ord}_2 = \{1, 2, 3, 4, 6\}$. We wish to state an explicit formula for Ord_N.

Define a function ψ based on Euler's totient function φ; it will map positive integers to non-negative integers. For an odd prime, p, and a positive integer, k, set $\psi(p^k)$ equal to the totient function value, $\varphi(p^k)$, which in this case is $p^k - p^{k-1}$. Do the same for $\psi(2^k)$ when $k > 1$. Set $\psi(2)$ and $\psi(1)$ to 0. Using the fundamental theorem of arithmetic, we can write any other positive integer uniquely as a product of prime powers, $m = \prod_\alpha p_\alpha^{k\alpha}$; set $\psi(m) = \sum_\alpha \psi(p_\alpha^{k\alpha})$. This differs from the totient itself, because it is a sum instead of a product.

The crystallographic restriction in general form states that Ord_N consists of those positive integers m such that $\psi(m) \le N$.

Smallest dimension for a given order																														
m	1	2	3	4	5	6	7	8	9	10	11	12	13	14	15	16	17	18	19	20	21	22	23	24	25	26	27	28	29	30	31
$\psi(m)$	0	0	2	2	4	2	6	4	6	4	10	4	12	6	6	8	16	6	18	6	8	10	22	6	20	12	18	8	28	6	30

For $m > 2$, the values of $\psi(m)$ are equal to twice the algebraic degree of $\cos(2\pi/m)$; therefore, $\psi(m)$ is strictly less than m and reaches this maximum value if and only if m is a prime.

Note that these additional symmetries do not allow a planar slice to have, say, 8-fold rotation symmetry. In the plane, the 2D restrictions still apply. Thus the cuts used to model quasicrystals necessarily have thickness.

Integer matrices are not limited to rotations; for example, a reflection is also a symmetry of order 2. But by insisting on determinant +1, we can restrict the matrices to proper rotations.

Formulation in Terms of Isometries

The crystallographic restriction theorem can be formulated in terms of isometries of Euclidean space. A set of isometries can form a group. By a *discrete isometry group* we will mean an isometry group that maps every point to a discrete subset of $\mathbb{R}^N$, i.e. a set of isolated points. With this terminology, the crystallographic restriction theorem in two and three dimensions can be formulated as follows.

For every discrete isometry group in two- and three-dimensional space which includes translations spanning the whole space, all isometries of finite order are of order 1, 2, 3, 4 or 6.

Note that isometries of order n include, but are not restricted to, n-fold rotations. The theorem also excludes S_8, S_{12}, D_{4d}, and D_{6d}, even though they have 4- and 6-fold rotational symmetry only.

Note also that rotational symmetry of any order about an axis is compatible with translational symmetry along that axis.

The result in the table above implies that for every discrete isometry group in four- and five-dimensional space which includes translations spanning the whole space, all isometries of finite order are of order 1, 2, 3, 4, 5, 6, 8, 10, or 12.

All isometries of finite order in six- and seven-dimensional space are of order 1, 2, 3, 4, 5, 6, 7, 8, 9, 10, 12, 14, 15, 18, 20, 24 or 30 .

X-ray Crystallography

X-ray crystallography (XRC) is a technique used to determine the atomic and molecular structure of a crystal, in which the crystalline structure causes a beam of incident X-rays to diffract into many specific directions. By measuring the angles and intensities of these diffracted beams, a crystallographer can produce a three-dimensional picture of the density of electrons within the crystal. From this electron density, the mean positions of the atoms in the crystal can be determined, as well as their chemical bonds, their crystallographic disorder, and various other information.

Since many materials can form crystals—such as salts, metals, minerals, semiconductors, as well as various inorganic, organic, and biological molecules—X-ray crystallography has been fundamental in the development of many scientific fields. In its first decades of use, this method determined the size of atoms, the lengths and types of chemical bonds, and the atomic-scale differences among various materials, especially minerals and alloys. The method also revealed the structure and function of many biological molecules, including vitamins, drugs, proteins and nucleic acids such as DNA. X-ray crystallography is still the primary method for characterizing the atomic structure of new materials and in discerning materials that appear similar by other experiments. X-ray crystal structures can also account for unusual electronic or elastic properties of a material, shed light on chemical interactions and processes, or serve as the basis for designing pharmaceuticals against diseases.

In a single-crystal X-ray diffraction measurement, a crystal is mounted on a goniometer. The goniometer is used to position the crystal at selected orientations. The crystal is illuminated with a finely focused monochromatic beam of X-rays, producing a diffraction pattern of regularly spaced spots known as reflections. The two-dimensional images taken at different orientations are converted into a three-dimensional model of the density of electrons within the crystal using the mathematical method of Fourier transforms, combined with chemical data known for the sample. Poor resolution (fuzziness) or even errors may result if the crystals are too small, or not uniform enough in their internal makeup.

X-ray crystallography is related to several other methods for determining atomic structures. Similar diffraction patterns can be produced by scattering electrons or neutrons, which are likewise interpreted by Fourier transformation. If single crystals of sufficient size cannot be obtained, various other X-ray methods can be applied to obtain less detailed information; such methods include fiber diffraction, powder diffraction and (if the sample is not crystallized) small-angle X-ray scattering (SAXS). If the material under investigation is only available in the form of nanocrystalline powders or suffers from poor crystallinity, the methods of electron crystallography can be applied for determining the atomic structure.

For all above mentioned X-ray diffraction methods, the scattering is elastic; the scattered X-rays have the same wavelength as the incoming X-ray. By contrast, inelastic X-ray scattering methods are useful in studying excitations of the sample such as plasmons, crystal-field and orbital excitations, magnons, and phonons, rather than the distribution of its atoms.

X-ray crystallography has led to a better understanding of chemical bonds and non-covalent interactions. The initial studies revealed the typical radii of atoms, and confirmed many theoretical models of chemical bonding, such as the tetrahedral bonding of carbon in the diamond structure,the octahedral bonding of metals observed in ammonium hexachloroplatinate (IV),and the resonance observed in the planar carbonate groupand in aromatic molecules. Kathleen Lonsdale's 1928 structure of hexamethyl benzene established the hexagonal symmetry of benzene and showed a clear difference in bond length between the aliphatic C–C bonds and aromatic C–C bonds; this finding led to the idea of resonance between chemical bonds, which had profound consequences for the development of chemistry. Her conclusions were anticipated by William Henry Bragg, who published models of naphthalene and anthracene in 1921 based on other molecules, an early form of molecular replacement.

Also in the 1920s, Victor Moritz Goldschmidt and later Linus Pauling developed rules for eliminating chemically unlikely structures and for determining the relative sizes of atoms. These rules led to the structure of brookite and an understanding of the relative stability of the rutile, brookite and anatase forms of titanium dioxide.

The distance between two bonded atoms is a sensitive measure of the bond strength and its bond order; thus, X-ray crystallographic studies have led to the discovery of even more exotic types of bonding in inorganic chemistry, such as metal-metal double bonds, metal-metal quadruple bonds, and three-center, two-electron bonds. X-ray crystallography—or, strictly speaking, an inelastic Compton scattering experiment—has also provided evidence for the partly covalent character of hydrogen bonds. In the field of organometallic chemistry, the X-ray structure of ferrocene initiated scientific studies of sandwich compounds, while that of Zeise's salt stimulated research into "back bonding" and metal-pi complexes. Finally, X-ray crystallography had a pioneering role in the development of supramolecular chemistry, particularly in clarifying the structures of the crown ethers and the principles of host–guest chemistry.

X-ray diffraction is a very powerful tool in catalyst development. Ex-situ measurements are carried out routinely for checking the crystal structure of materials or to unravel new structures. In-situ experiments give comprehensive understanding about the structural stability of catalysts under reaction conditions.

In material sciences, many complicated inorganic and organometallic systems have been analyzed using single-crystal methods, such as fullerenes, metalloporphyrins, and other complicated compounds. Single-crystal diffraction is also used in the pharmaceutical industry, due to recent problems with polymorphs. The major factors affecting the quality of single-crystal structures are the crystal's size and regularity; recrystallization is a commonly used technique to improve these factors in small-molecule crystals. The Cambridge Structural Database contains over 1,000,000 structures as of June 2019; over 99% of these structures were determined by X-ray diffraction.

Mineralogy and Metallurgy

First X-ray diffraction view of Martian soil – CheMin analysis reveals feldspar, pyroxenes, olivine and more.

Since the 1920s, X-ray diffraction has been the principal method for determining the arrangement of atoms in minerals and metals. The application of X-ray crystallography to mineralogy began with the structure of garnet, which was determined in 1924 by Menzer. A systematic X-ray crystallographic study of the silicates was undertaken in the 1920s. This study showed that, as the Si/O ratio is altered, the silicate crystals exhibit significant changes in their atomic arrangements. Machatschki extended these insights to minerals in which aluminium substitutes for the silicon atoms of the silicates. The first application of X-ray crystallography to metallurgy likewise occurred in the mid-1920s. Most notably, Linus Pauling's structure of the alloy Mg_2Sn led to his theory of the stability and structure of complex ionic crystals.

On October 17, 2012, the Curiosity rover on the planet Mars at "Rocknest" performed the first X-ray diffraction analysis of Martian soil. The results from the rover's CheMin analyzer revealed the presence of several minerals, including feldspar, pyroxenes and olivine, and suggested that the Martian soil in the sample was similar to the "weathered basaltic soils" of Hawaiian volcanoes.

Early Organic and Small Biological Molecules

The first structure of an organic compound, hexamethyl enetetramine, was solved in 1923. This was followed by several studies of long-chain fatty acids, which are an important component of biological membranes. In the 1930s, the structures of much larger molecules with two-dimensional complexity began to be solved. A significant advance was the structure of phthalocyanine, a large planar molecule that is closely related to porphyrin molecules important in biology, such as heme, corrin and chlorophyll.

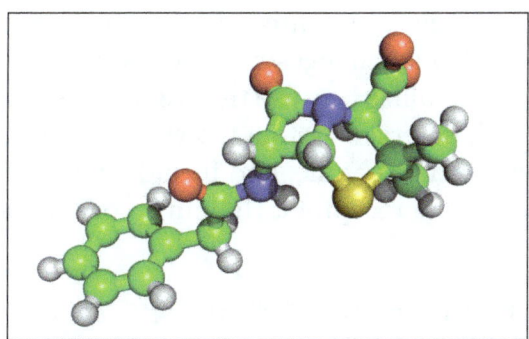

The three-dimensional structure of penicillin, solved by Dorothy Crowfoot Hodgkin in 1945. The green, red, yellow and blue spheres represent atoms of carbon, oxygen, sulfur and nitrogen, respectively. The white spheres represent hydrogen, which were determined mathematically rather than by the X-ray analysis.

X-ray crystallography of biological molecules took off with Dorothy Crowfoot Hodgkin, who solved the structures of cholesterol, penicillin and vitamin B_{12}, for which she was awarded the Nobel Prize in Chemistry in 1964.

Biological Macromolecular Crystallography

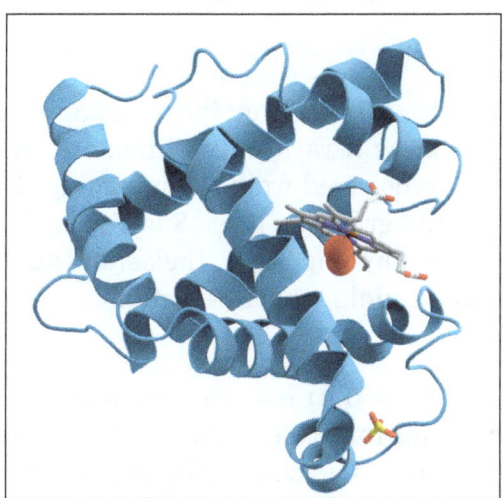

Ribbon diagram of the structure of myoglobin, showing colored alpha helices. Such proteins are long, linear molecules with thousands of atoms; yet the relative position of each atom has been determined with sub-atomic resolution by X-ray crystallography. Since it is difficult to visualize all the atoms at once, the ribbon shows the rough path of the protein polymer from its N-terminus (blue) to its C-terminus (red).

Crystal structures of proteins (which are irregular and hundreds of times larger than cholesterol) began to be solved in the late 1950s, beginning with the structure of sperm whale myoglobin by Sir John Cowdery Kendrew, for which he shared the Nobel Prize in Chemistry with Max Perutz in 1962. Since that success, 132055 X-ray crystal structures of proteins, nucleic acids and other biological molecules have been determined. For comparison, the nearest competing method in terms of structures analyzed is nuclear magnetic resonance (NMR) spectroscopy, which has resolved 11904 chemical structures. Moreover, crystallography can solve structures of arbitrarily large molecules, whereas solution-state NMR is restricted to relatively small ones (less than 70 kDa). X-ray crystallography is now used routinely by scientists to determine how a pharmaceutical drug interacts with its protein target and what changes might improve it. However, intrinsic membrane proteins remain challenging to crystallize because they require detergents or other means to solubilize them in isolation, and such detergents often interfere with crystallization. Such membrane proteins are a large component of the genome, and include many proteins of great physiological importance, such as ion channels and receptors. Helium cryogenics are used to prevent radiation damage in protein crystals.

On the other end of the size scale, even relatively small molecules may pose challenges for the resolving power of X-ray crystallography. The structure assigned in 1991 to the antibiotic isolated from a marine organism, diazonamide A ($C_{40}H_{34}Cl_2N_6O_6$, molar mass 765.65 g/mol), proved to be incorrect by the classical proof of structure: a synthetic sample was not identical to the natural product. The mistake was attributed to the inability of X-ray crystallography to distinguish between the correct -OH / >NH and the interchanged $-NH_2$ / -O- groups in the incorrect structure. With advances in instrumentation, however, it is now routinely possible to distinguish between such similar groups using modern single-crystal X-ray diffractometers.

Relationship to other Scattering Techniques

Elastic vs. Inelastic Scattering

X-ray crystallography is a form of elastic scattering; the outgoing X-rays have the same energy, and thus same wavelength, as the incoming X-rays, only with altered direction. By contrast, inelastic scattering occurs when energy is transferred from the incoming X-ray to the crystal, e.g., by exciting an inner-shell electron to a higher energy level. Such inelastic scattering reduces the energy (or increases the wavelength) of the outgoing beam. Inelastic scattering is useful for probing such excitations of matter, but not in determining the distribution of scatterers within the matter, which is the goal of X-ray crystallography.

X-rays range in wavelength from 10 to 0.01 nanometers; a typical wavelength used for crystallography is 1 Å (0.1 nm),which is on the scale of covalent chemical bonds and the radius of a single atom. Longer-wavelength photons (such as ultraviolet radiation) would not have sufficient resolution to determine the atomic positions. At the other extreme, shorter-wavelength photons such as gamma rays are difficult to produce in large numbers, difficult to focus, and interact too strongly with matter, producing particle-antiparticle pairs. Therefore, X-rays are the "sweetspot" for wavelength when determining atomic-resolution structures from the scattering of electromagnetic radiation.

Other X-ray Techniques

Other forms of elastic X-ray scattering include powder diffraction, Small-Angle X-ray Scattering (SAXS) and several types of X-ray fiber diffraction, which was used by Rosalind Franklin in determining the double-helix structure of DNA. In general, single-crystal X-ray diffraction offers more structural information than these other techniques; however, it requires a sufficiently large and regular crystal, which is not always available.

These scattering methods generally use monochromatic X-rays, which are restricted to a single wavelength with minor deviations. A broad spectrum of X-rays (that is, a blend of X-rays with different wavelengths) can also be used to carry out X-ray diffraction, a technique known as the Laue method. This is the method used in the original discovery of X-ray diffraction. Laue scattering provides much structural information with only a short exposure to the X-ray beam, and is therefore used in structural studies of very rapid events (Time resolved crystallography). However, it is not as well-suited as monochromatic scattering for determining the full atomic structure of a crystal and therefore works better with crystals with relatively simple atomic arrangements.

The Laue back reflection mode records X-rays scattered backwards from a broad spectrum source. This is useful if the sample is too thick for X-rays to transmit through it. The diffracting planes in the crystal are determined by knowing that the normal to the diffracting plane bisects the angle between the incident beam and the diffracted beam. A Greninger chart can be used to interpret the back reflection Laue photograph.

Electron and Neutron Diffraction

Other particles, such as electrons and neutrons, may be used to produce a diffraction pattern. Although electron, neutron, and X-ray scattering are based on different physical processes, the resulting diffraction patterns are analyzed using the same coherent diffraction imaging techniques.

As derived below, the electron density within the crystal and the diffraction patterns are related by a simple mathematical method, the Fourier transform, which allows the density to be calculated relatively easily from the patterns. However, this works only if the scattering is *weak*, i.e., if the scattered beams are much less intense than the incoming beam. Weakly scattered beams pass through the remainder of the crystal without undergoing a second scattering event. Such re-scattered waves are called "secondary scattering" and hinder the analysis. Any sufficiently thick crystal will produce secondary scattering, but since X-rays interact relatively weakly with the electrons, this is generally not a significant concern. By contrast, electron beams may produce strong secondary scattering even for relatively thin crystals (>100 nm). Since this thickness corresponds to the diameter of many viruses, a promising direction is the electron diffraction of isolated macromolecular assemblies, such as viral capsids and molecular machines, which may be carried out with a cryo-electron microscope. Moreover, the strong interaction of electrons with matter (about 1000 times stronger than for X-rays) allows determination of the atomic structure of extremely small volumes. The field of applications for electron crystallography ranges from bio molecules like membrane proteins over organic thin films to the complex structures of (nanocrystalline) intermetallic compounds and zeolites.

Neutron diffraction is an excellent method for structure determination, although it has been difficult to obtain intense, monochromatic beams of neutrons in sufficient quantities. Traditionally, nuclear reactors have been used, although sources producing neutrons by spallation are becoming increasingly available. Being uncharged, neutrons scatter much more readily from the atomic nuclei rather than from the electrons. Therefore, neutron scattering is very useful for observing the positions of light atoms with few electrons, especially hydrogen, which is essentially invisible in the X-ray diffraction. Neutron scattering also has the remarkable property that the solvent can be made invisible by adjusting the ratio of normal water, H_2O, and heavy water, D_2O.

Methods

Overview of Single-crystal X-ray Diffraction

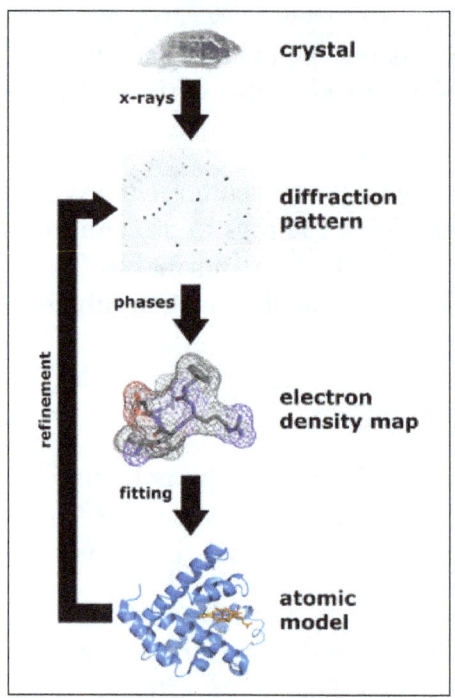

Workflow for solving the structure of a molecule by X-ray crystallography.

The oldest and most precise method of X-ray crystallography is single-crystal X-ray diffraction, in which a beam of X-rays strikes a single crystal, producing scattered beams. When they land on a piece of film or other detector, these beams make a diffraction pattern of spots; the strengths and angles of these beams are recorded as the crystal is gradually rotated. Each spot is called a reflection, since it corresponds to the reflection of the X-rays from one set of evenly spaced planes within the crystal. For single crystals of sufficient purity and regularity, X-ray diffraction data can determine the mean chemical bond lengths and angles to within a few thousandths of an angstrom and to within a few tenths of a degree, respectively. The atoms in a crystal are not static, but oscillate about their mean positions, usually by less than a few tenths of an angstrom. X-ray crystallography allows measuring the size of these oscillations.

Procedure

The technique of single-crystal X-ray crystallography has three basic steps. The first—and often most difficult—step is to obtain an adequate crystal of the material under study. The crystal should be sufficiently large (typically larger than 0.1 mm in all dimensions), pure in composition and regular in structure, with no significant internal imperfections such as cracks or twinning.

In the second step, the crystal is placed in an intense beam of X-rays, usually of a single wavelength (*monochromatic X-rays*), producing the regular pattern of reflections. The angles and intensities of diffracted X-rays are measured, with each compound having a unique diffraction pattern.As the crystal is gradually rotated, previous reflections disappear and new ones appear; the intensity of every spot is recorded at every orientation of the crystal. Multiple data sets may have to be collected, with each set covering slightly more than half a full rotation of the crystal and typically containing tens of thousands of reflections.

In the third step, these data are combined computationally with complementary chemical information to produce and refine a model of the arrangement of atoms within the crystal. The final, refined model of the atomic arrangement—now called a crystal structure—is usually stored in a public database.

Limitations

As the crystal's repeating unit, its unit cell, becomes larger and more complex, the atomic-level picture provided by X-ray crystallography becomes less well-resolved (more "fuzzy") for a given number of observed reflections. Two limiting cases of X-ray crystallography—"small-molecule" (which includes continuous inorganic solids) and "macromolecular" crystallography—are often discerned. Small-molecule crystallography typically involves crystals with fewer than 100 atoms in their asymmetric unit; such crystal structures are usually so well resolved that the atoms can be discerned as isolated "blobs" of electron density. By contrast, macromolecular crystallography often involves tens of thousands of atoms in the unit cell. Such crystal structures are generally less well-resolved (more "smeared out"); the atoms and chemical bonds appear as tubes of electron density, rather than as isolated atoms. In general, small molecules are also easier to crystallize than macromolecules; however, X-ray crystallography has proven possible even for viruses and proteins with hundreds of thousands of atoms, through improved crystallographic imaging and technology. Though normally X-ray crystallography can only be performed if the sample is in crystal form, new research has been done into sampling non-crystalline forms of samples.

Crystallization

A protein crystal seen under a microscope. Crystals used in X-ray crystallography
may be smaller than a millimeter across.

Although crystallography can be used to characterize the disorder in an impure or irregular crystal, crystallography generally requires a pure crystal of high regularity to solve the structure of a complicated arrangement of atoms. Pure, regular crystals can sometimes be obtained from natural or synthetic materials, such as samples of metals, minerals or other macroscopic materials. The regularity of such crystals can sometimes be improved with macromolecular crystal annealing and other methods. However, in many cases, obtaining a diffraction-quality crystal is the chief barrier to solving its atomic-resolution structure.

Small-molecule and macromolecular crystallography differ in the range of possible techniques used to produce diffraction-quality crystals. Small molecules generally have few degrees of conformational freedom, and may be crystallized by a wide range of methods, such as chemical vapor deposition and recrystallization. By contrast, macromolecules generally have many degrees of freedom and their crystallization must be carried out so as to maintain a stable structure. For example, proteins and larger RNA molecules cannot be crystallized if their tertiary structure has been unfolded; therefore, the range of crystallization conditions is restricted to solution conditions in which such molecules remain folded.

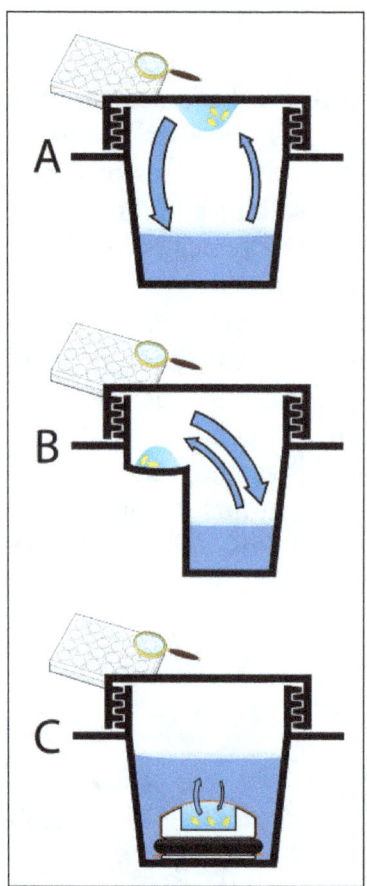

Three methods of preparing crystals, A: Hanging drop. B: Sitting drop. C: Microdialysis.

Protein crystals are almost always grown in solution. The most common approach is to lower the solubility of its component molecules very gradually; if this is done too quickly, the molecules will precipitate from solution, forming a useless dust or amorphous gel on the bottom of the container. Crystal growth in solution is characterized by two steps: nucleation of a microscopic crystallite (possibly having only 100 molecules), followed by growth of that crystallite, ideally to a diffraction-quality crystal. The solution conditions that favor the first step (nucleation) are not always the same conditions that favor the second step (subsequent growth). The crystallographer's goal is to identify solution conditions that favor the development of a single, large crystal, since larger crystals offer improved resolution of the molecule. Consequently, the solution conditions should disfavor the first step (nucleation) but favor the second (growth), so that only one large crystal forms per droplet. If nucleation is favored too much, a shower of small crystallites will form in the droplet, rather than one large crystal; if favored too little, no crystal will form whatsoever. Other approaches involves, crystallizing proteins under oil, where aqueous protein solutions are dispensed under liquid oil, and water evaporates through the layer of oil. Different oils have different evaporation permeabilities, therefore yielding changes in concentration rates from different percipient/protein mixture.

It is extremely difficult to predict good conditions for nucleation or growth of well-ordered crystals. In practice, favorable conditions are identified by screening; a very large batch of the molecules is prepared, and a wide variety of crystallization solutions are tested. Hundreds, even thousands, of solution conditions are generally tried before finding the successful one. The various conditions can use one or more physical mechanisms to lower the solubility of the molecule; for example, some may change the pH, some contain salts of the Hofmeister series or chemicals that lower the dielectric constant of the solution, and still others contain large polymers such as polyethylene glycol that drive the molecule out of solution by entropic effects. It is also common to try several temperatures for encouraging crystallization, or to gradually lower the temperature so that the solution becomes supersaturated. These methods require large amounts of the target molecule, as they use high concentration of the molecule(s) to be crystallized. Due to the difficulty in obtaining such large quantities (milligrams) of crystallization-grade protein, robots have been developed that are capable of accurately dispensing crystallization trial drops that are in the order of 100 nanoliters in volume. This means that 10-fold less protein is used per experiment when compared to crystallization trials set up by hand (in the order of 1 microliter).

Several factors are known to inhibit or mar crystallization. The growing crystals are generally held at a constant temperature and protected from shocks or vibrations that might disturb their crystallization. Impurities in the molecules or in the crystallization solutions are often inimical to crystallization. Conformational flexibility in the molecule also tends to make crystallization less likely, due to entropy. Molecules that tend to self-assemble into regular helices are often unwilling to assemble into crystals. Crystals can be marred by twinning, which can occur when a unit cell can pack equally favorably in multiple orientations; although recent advances in computational methods may allow solving the structure of some twinned crystals. Having failed to crystallize a target molecule, a crystallographer may try again with a slightly modified version of the molecule; even small changes in molecular properties can lead to large differences in crystallization behavior.

Data Collection

Mounting the Crystal

The crystal is mounted for measurements so that it may be held in the X-ray beam and rotated. There are several methods of mounting. In the past, crystals were loaded into glass capillaries with the crystallization solution (the mother liquor). Nowadays, crystals of small molecules are typically attached with oil or glue to a glass fiber or a loop, which is made of nylon or plastic and attached to a solid rod. Protein crystals are scooped up by a loop, then flash-frozen with liquid nitrogen. This freezing reduces the radiation damage of the X-rays, as well as the noise in the Bragg peaks due to thermal motion (the Debye-Waller effect). However, untreated protein crystals often crack if flash-frozen; therefore, they are generally pre-soaked in a cryoprotectant solution before freezing. Unfortunately, this pre-soak may itself cause the crystal to crack, ruining it for crystallography. Generally, successful cryo-conditions are identified by trial and error.

The capillary or loop is mounted on a goniometer, which allows it to be positioned accurately within the X-ray beam and rotated. Since both the crystal and the beam are often very small, the crystal must be centered within the beam to within ~25 micrometers accuracy, which is aided by a camera focused on the crystal. The most common type of goniometer is the "kappa goniometer", which offers three angles of rotation: the ω angle, which rotates about an axis perpendicular to the beam; the κ angle, about an axis at ~50° to the ω axis; and, finally, the φ angle about the loop/capillary axis. When the κ angle is zero, the ω and φ axes are aligned. The κ rotation allows for convenient mounting of the crystal, since the arm in which the crystal is mounted may be swung out towards the crystallographer. The oscillations carried out during data collection (mentioned below) involve the ω axis only. An older type of goniometer is the four-circle goniometer, and its relatives such as the six-circle goniometer.

X-ray Sources

Rotating Anode

Small scale can be done on a local X-ray tube source, typically coupled with an image plate detector. These have the advantage of being (relatively) inexpensive and easy to maintain, and allow for quick screening and collection of samples. However, the wavelength light produced is limited by anode material, typically copper. Further, intensity is limited by the power applied and cooling capacity available to avoid melting the anode. In such systems, electrons are boiled off of a cathode and accelerated through a strong electric potential of ~50 kV; having reached a high speed, the electrons collide with a metal plate, emitting *bremsstrahlung* and some strong spectral lines corresponding to the excitation of inner-shell electrons of the metal. The most common metal used is copper, which can be kept cool easily, due to its high thermal conductivity, and which produces strong K_α and K_β lines. The K_β line is sometimes suppressed with a thin (~10 μm) nickel foil. The simplest and cheapest variety of sealed X-ray tube has a stationary anode (the Crookes tube) and run with ~2 kW of electron beam power. The more expensive variety has a rotating-anode type source that run with ~14 kW of e-beam power.

X-rays are generally filtered (by use of X-ray filters) to a single wavelength (made monochromatic) and collimated to a single direction before they are allowed to strike the crystal. The filtering not only simplifies the data analysis, but also removes radiation that degrades the crystal without contributing useful information. Collimation is done either with a collimator (basically, a long tube) or with a clever arrangement of gently curved mirrors. Mirror systems are preferred for small crystals (under 0.3 mm) or with large unit cells (over 150 Å).

Rotating anodes were used by Joanna Maria Vandenberg in the first experiments that demonstrated the power of X rays for quick (in real time production) screening of large InGaAsP thin film wafers for quality control of quantum well lasers.

Synchrotron Radiation

Synchrotron radiation sources are some of the brightest lights on earth. It is the single most powerful tool available to X-ray crystallographers. It is made of X-ray beams generated in large machines called synchrotrons. These machines accelerate electrically charged particles, often electrons, to nearly the speed of light and confine them in a (roughly) circular loop using magnetic fields.

Synchrotrons are generally national facilities, each with several dedicated beamlines where data is collected without interruption. Synchrotrons were originally designed for use by high-energy physicists studying subatomic particles and cosmic phenomena. The largest component of each synchrotron is its electron storage ring. This ring is actually not a perfect circle, but a many-sided polygon. At each corner of the polygon, or sector, precisely aligned magnets bend the electron stream. As the electrons' path is bent, they emit bursts of energy in the form of X-rays.

Using synchrotron radiation frequently has specific requirements for X-ray crystallography. The intense ionizing radiation can cause radiation damage to samples, particularly macromolecular crystals. Cryo crystallography protects the sample from radiation damage, by freezing the crystal at liquid nitrogen temperatures (~100 K). However, synchrotron radiation frequently has the advantage of user selectable wavelengths, allowing for anomalous scattering experiments which maximizes anomalous signal. This is critical in experiments such as SAD and MAD.

Free Electron Laser

Recently, free-electron lasers have been developed for use in X-ray crystallography. These are the brightest X-ray sources currently available; with the X-rays coming in femtosecond bursts. The intensity of the source is such that atomic resolution diffraction patterns can be resolved for crystals otherwise too small for collection. However, the intense light source also destroys the sample, requiring multiple crystals to be shot. As each crystal is randomly oriented in the beam, hundreds of thousands of individual diffraction images must be collected in order to get a complete data-set. This method, serial femtosecond crystallography, has been used in solving the structure of a number of protein crystal structures, sometimes noting differences with equivalent structures collected from synchrotron sources.

Recording the Reflections

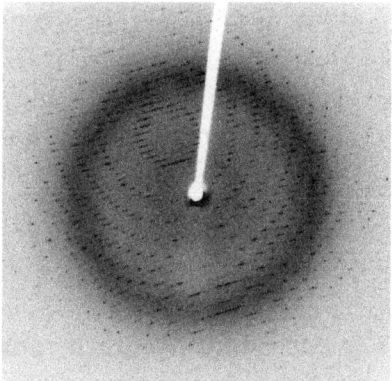

An X-ray diffraction pattern of a crystallized enzyme. The pattern of spots (reflections) and the relative strength of each spot (intensities) can be used to determine the structure of the enzyme.

When a crystal is mounted and exposed to an intense beam of X-rays, it scatters the X-rays into a pattern of spots or reflections that can be observed on a screen behind the crystal. A similar pattern may be seen by shining a laser pointer at a compact disc. The relative intensities of these spots provide the information to determine the arrangement of molecules within the crystal in atomic detail. The intensities of these reflections may be recorded with photographic film, an area detector (such as a pixel detector) or with a charge-coupled device (CCD) image sensor. The peaks at small angles correspond to low-resolution data, whereas those at high angles represent high-resolution data; thus, an upper limit on the eventual resolution of the structure can be determined from the first few images. Some measures of diffraction quality can be determined at this point, such as the mosaicity of the crystal and its overall disorder, as observed in the peak widths. Some pathologies of the crystal that would render it unfit for solving the structure can also be diagnosed quickly at this point.

One image of spots is insufficient to reconstruct the whole crystal; it represents only a small slice of the full Fourier transform. To collect all the necessary information, the crystal must be rotated step-by-step through 180°, with an image recorded at every step; actually, slightly more than 180° is required to cover reciprocal space, due to the curvature of the Ewald sphere. However, if the crystal has a higher symmetry, a smaller angular range such as 90° or 45° may be recorded. The rotation axis should be changed at least once, to avoid developing a "blind spot" in reciprocal space close to the rotation axis. It is customary to rock the crystal slightly (by 0.5–2°) to catch a broader region of reciprocal space.

Multiple data sets may be necessary for certain phasing methods. For example, MAD phasing requires that the scattering be recorded at least three (and usually four, for redundancy) wavelengths of the incoming X-ray radiation. A single crystal may degrade too much during the collection of one data set, owing to radiation damage; in such cases, data sets on multiple crystals must be taken.

Data Analysis

Crystal Symmetry, Unit Cell and Image Scaling

The recorded series of two-dimensional diffraction patterns, each corresponding to a different crystal orientation, is converted into a three-dimensional model of the electron density; the conversion uses the mathematical technique of Fourier transforms, which is explained below. Each spot corresponds to a different type of variation in the electron density; the crystallographer must determine which variation corresponds to which spot (indexing), the relative strengths of the spots in different images (merging and scaling) and how the variations should be combined to yield the total electron density (phasing).

Data processing begins with *indexing* the reflections. This means identifying the dimensions of the unit cell and which image peak corresponds to which position in reciprocal space. A byproduct of indexing is to determine the symmetry of the crystal, i.e., its *space group*. Some space groups can be eliminated from the beginning. For example, reflection symmetries cannot be observed in chiral molecules; thus, only 65 space groups of 230 possible are allowed for protein molecules which are almost always chiral. Indexing is generally accomplished using an *autoindexing* routine. Having assigned symmetry, the data is then *integrated*. This converts the hundreds of images containing the thousands of reflections into a single file, consisting of (at the very least) records of the Miller index of each reflection, and an intensity for each reflection (at this state the file often also includes error estimates and measures of partiality (what part of a given reflection was recorded on that image)).

A full data set may consist of hundreds of separate images taken at different orientations of the crystal. The first step is to merge and scale these various images, that is, to identify which peaks appear in two or more images (*merging*) and to scale the relative images so that they have a consistent intensity scale. Optimizing the intensity scale is critical because the relative intensity of the peaks is the key information from which the structure is determined. The repetitive technique of crystallographic data collection and the often high symmetry of crystalline materials cause the diffractometer to record many symmetry-equivalent reflections multiple times. This allows calculating the symmetry-related R-factor, a reliability index based upon how similar are the measured intensities of symmetry-equivalent reflections, thus assessing the quality of the data.

Initial Phasing

The data collected from a diffraction experiment is a reciprocal space representation of the crystal lattice. The position of each diffraction 'spot' is governed by the size and shape of the unit cell, and the inherent symmetry within the crystal. The intensity of each diffraction 'spot' is recorded, and this intensity is proportional to the square of the *structure factor* amplitude. The structure factor is a complex number containing information relating to both the amplitude and phase of a wave. In order to obtain an interpretable *electron density map*, both amplitude and phase must be known (an electron density map allows a crystallographer to build a starting model of the molecule). The phase cannot be directly recorded during a diffraction experiment: this is known as the phase problem. Initial phase estimates can be obtained in a variety of ways:

- Ab initio phasing or direct methods – This is usually the method of choice for small molecules (<1000 non-hydrogen atoms), and has been used successfully to solve the phase problems for small proteins. If the resolution of the data is better than 1.4 Å (140 pm), direct methods can be used to obtain phase information, by exploiting known phase relationships between certain groups of reflections.

- Molecular replacement – If a related structure is known, it can be used as a search model in molecular replacement to determine the orientation and position of the molecules within the unit cell. The phases obtained this way can be used to generate *electron density maps*.

- Anomalous X-ray scattering (*MAD or SAD phasing*) – The X-ray wavelength may be scanned past an absorption edge of an atom, which changes the scattering in a known way. By recording full sets of reflections at three different wavelengths (far below, far above and in the middle of the absorption edge) one can solve for the substructure of the anomalously diffracting atoms and hence the structure of the whole molecule. The most popular method of incorporating anomalous scattering atoms into proteins is to express the protein in a methionine auxotroph (a host incapable of synthesizing methionine) in a media rich in seleno-methionine, which contains selenium atoms. A MAD experiment can then be conducted around the absorption edge, which should then yield the position of any methionine residues within the protein, providing initial phases.

- Heavy atom methods (multiple isomorphous replacement) – If electron-dense metal atoms can be introduced into the crystal, direct methods or Patterson-space methods can be used to determine their location and to obtain initial phases. Such heavy atoms can be introduced either by soaking the crystal in a heavy atom-containing solution, or by co-crystallization (growing the crystals in the presence of a heavy atom). As in MAD phasing, the changes in the scattering amplitudes can be interpreted to yield the phases. Although this is the original method by which protein crystal structures were solved, it has largely been superseded by MAD phasing with selenomethionine.

Model Building and Phase Refinement

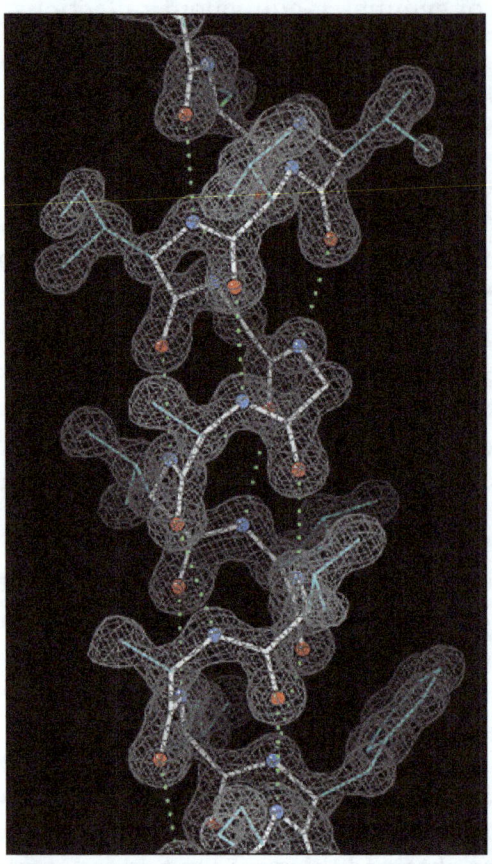

Structure of a protein alpha helix, with stick-figures for the covalent bonding within electron density for the crystal structure at ultra-high-resolution (0.91 Å). The density contours are in gray, the helix backbone in white, sidechains in cyan, O atoms in red, N atoms in blue, and hydrogen bonds as green dotted lines.

Having obtained initial phases, an initial model can be built. The atomic positions in the model and their respective Debye-Waller factors (or B-factors, accounting for the thermal motion of the atom) can be refined to fit the observed diffraction data, ideally yielding a better set of phases. A new model can then be fit to the new electron density map and successive rounds of refinement is carried out. This interative process continues until the correlation between the diffraction data and the model is maximized. The agreement is measured by an R-factor defined as:

$$R = \frac{\displaystyle\sum_{\text{all reflections}} \left| F_{\text{obs}} - F_{\text{calc}} \right|}{\displaystyle\sum_{\text{all reflections}} \left| F_{\text{obs}} \right|},$$

Here F is the structure factor. A similar quality criterion is R_{free}, which is calculated from a subset (~10%) of reflections that were not included in the structure refinement. Both R factors depend on the resolution of the data. As a rule of thumb, R_{free} should be approximately the resolution in angstroms divided by 10; thus, a data-set with 2 Å resolution should yield a final R_{free} ~ 0.2. Chemical bonding features such as stereochemistry, hydrogen bonding and distribution of bond lengths and angles are complementary measures of the model quality. Phase bias is a serious problem in such iterative model building. *Omit maps* are a common technique used to check for this.

It may not be possible to observe every atom in the asymmetric unit. In many cases, Crystallographic disorder smears the electron density map. Weakly scattering atoms such as hydrogen are routinely invisible. It is also possible for a single atom to appear multiple times in an electron density map, e.g., if a protein sidechain has multiple (<4) allowed conformations. In still other cases, the crystallographer may detect that the covalent structure deduced for the molecule was incorrect, or changed. For example, proteins may be cleaved or undergo post-translational modifications that were not detected prior to the crystallization.

Disorder

A common challenge in refinement of crystal structures results from crystallographic disorder. Disorder can take many forms but in general involves the coexistence of two or more species or conformations. Failure to recognize disorder results in flawed interpretation. Pitfalls from improper modeling of disorder are illustrated by the discounted hypothesis of bond stretch isomerism. Disorder is modelled with respect to the relative population of the components, often only two, and their identity. In structures of large molecules and ions, solvent and counterions are often disordered.

Applied Computational Data Analysis

The use of computational methods for the powder X-ray diffraction data analysis is now generalized. It typically compares the experimental data to the simulated diffractogram of a model structure, taking into account the instrumental parameters, and refines the structural or microstructural parameters of the model using least squares based minimization algorithm. Most available tools allowing phase identification and structural refinement are based on the Rietveld method, some of them being open and free software such as FullProf Suite, Jana 2006, MAUD, Rietan, GSAS, etc. Most of these tools also allow Le Bail refinement (also referred to as profile matching), that is, refinement of the cell parameters based on the Bragg peaks positions and peak profiles, without taking into account the crystallographic structure by itself. More recent tools allow the refinement of both structural and microstructural data, such as the FAULTS program included in the FullProf Suite, which allows the refinement of structures with planar defects (e.g. stacking faults, twinnings, intergrowths).

Deposition of the Structure

Once the model of a molecule's structure has been finalized, it is often deposited in a crystallographic database such as the Cambridge Structural Database (for small molecules), the Inorganic Crystal Structure Database (ICSD) (for inorganic compounds) or the Protein Data Bank (for protein and sometimes nucleic acids). Many structures obtained in private commercial ventures to crystallize medicinally relevant proteins are not deposited in public crystallographic databases.

Diffraction Theory

The main goal of X-ray crystallography is to determine the density of electrons $f(r)$ throughout the crystal, where r represents the three-dimensional position vector within the crystal. To do this, X-ray scattering is used to collect data about its Fourier transform $F(q)$, which is inverted mathematically to obtain the density defined in real space, using the formula:

$$f(\mathrm{r}) = \frac{1}{(2\pi)^3} \int F(\mathrm{q}) e^{i\mathrm{q}\cdot\mathrm{r}} d\mathrm{q},$$

where the integral is taken over all values of q. The three-dimensional real vector q represents a point in reciprocal space, that is, to a particular oscillation in the electron density as one moves in the direction in which q points. The length of q corresponds to 2π divided by the wavelength of the oscillation. The corresponding formula for a Fourier transform will be used below:

$$F(\mathrm{q}) = \int f(\mathrm{r}) e^{-i\mathrm{q}\cdot\mathrm{r}} d\mathrm{r},$$

where the integral is summed over all possible values of the position vector r within the crystal.

The Fourier transform $F(q)$ is generally a complex number, and therefore has a magnitude $|F(q)|$ and a phase $\varphi(q)$ related by the equation

$$F(\mathrm{q}) = \left|F(\mathrm{q})\right| e^{i\phi(\mathrm{q})}.$$

The intensities of the reflections observed in X-ray diffraction give us the magnitudes $|F(q)|$ but not the phases $\varphi(q)$. To obtain the phases, full sets of reflections are collected with known alterations to the scattering, either by modulating the wavelength past a certain absorption edge or by adding strongly scattering (i.e., electron-dense) metal atoms such as mercury. Combining the magnitudes and phases yields the full Fourier transform $F(q)$, which may be inverted to obtain the electron density $f(r)$.

Crystals are often idealized as being *perfectly* periodic. In that ideal case, the atoms are positioned on a perfect lattice, the electron density is perfectly periodic, and the Fourier transform $F(q)$ is zero except when q belongs to the reciprocal lattice (the so-called *Bragg peaks*). In reality, however, crystals are not perfectly periodic; atoms vibrate about their mean position, and there may be disorder of various types, such as mosaicity, dislocations, various point defects, and heterogeneity in the conformation of crystallized molecules. Therefore, the Bragg peaks have a finite width and there may be significant *diffuse scattering*, a continuum of scattered X-rays that fall between the Bragg peaks.

Intuitive Understanding by Bragg's Law

An intuitive understanding of X-ray diffraction can be obtained from the Bragg model of diffraction. In this model, a given reflection is associated with a set of evenly spaced sheets running through the crystal, usually passing through the centers of the atoms of the crystal lattice. The orientation of a particular set of sheets is identified by its three Miller indices (h, k, l), and let their spacing be noted by d. William Lawrence Bragg proposed a model in which the incoming X-rays are scattered specularly (mirror-like) from each plane; from that assumption, X-rays scattered from adjacent planes will combine constructively (constructive interference) when the angle θ between the plane and the X-ray results in a path-length difference that is an integer multiple n of the X-ray wavelength λ.

$$2d \sin \theta = n\lambda.$$

A reflection is said to be *indexed* when its Miller indices (or, more correctly, its reciprocal lattice vector components) have been identified from the known wavelength and the scattering angle 2θ. Such indexing gives the unit-cell parameters, the lengths and angles of the unit-cell, as well as its space group. Since Bragg's law does not interpret the relative intensities of the reflections, however, it is generally inadequate to solve for the arrangement of atoms within the unit-cell; for that, a Fourier transform method must be carried out.

Scattering as a Fourier Transform

The incoming X-ray beam has a polarization and should be represented as a vector wave; however, for simplicity, let it be represented here as a scalar wave. We also ignore the complication of the time dependence of the wave and just concentrate on the wave's spatial dependence. Plane waves can be represented by a wave vector k_{in}, and so the strength of the incoming wave at time $t = 0$ is given by:

$$A e^{i k_{in} \cdot r}.$$

At position r within the sample, let there be a density of scatterers $f(r)$; these scatterers should produce a scattered spherical wave of amplitude proportional to the local amplitude of the incoming wave times the number of scatterers in a small volume dV about r.

$$\text{Amplitude of scattered wave} = A e^{i k_{in} \cdot r} S f(r) dV,$$

where S is the proportionality constant.

Consider the fraction of scattered waves that leave with an outgoing wave-vector of k_{out} and strike the screen at r_{screen}. Since no energy is lost (elastic, not inelastic scattering), the wavelengths are the same as are the magnitudes of the wave-vectors $|k_{in}| = |k_{out}|$. From the time that the photon is scattered at r until it is absorbed at r_{screen}, the photon undergoes a change in phase:

$$e^{i k_{out} \cdot (r_{screen} - r)}.$$

The net radiation arriving at r_{screen} is the sum of all the scattered waves throughout the crystal:

$$A S \int dr f(r) e^{i k_{in} \cdot r} e^{i k_{out} \cdot (r_{screen} - r)} = A S e^{i k_{out} \cdot r_{screen}} \int dr f(r) e^{i(k_{in} - k_{out}) \cdot r},$$

which may be written as a Fourier transform

$$A S e^{i k_{out} \cdot r_{screen}} \int dr f(r) e^{-i q \cdot r} = A S e^{i k_{out} \cdot r_{screen}} F(q),$$

where $q = k_{out} - k_{in}$. The measured intensity of the reflection will be square of this amplitude

$$A^2 S^2 |F(q)|^2.$$

Friedel and Bijvoet Mates

For every reflection corresponding to a point q in the reciprocal space, there is another reflection of the same intensity at the opposite point -q. This opposite reflection is known as the *Friedel mate* of the original reflection. This symmetry results from the mathematical fact that the density of electrons $f(r)$ at a position r is always a real number. As noted above, $f(r)$ is the inverse transform of its Fourier transform $F(q)$; however, such an inverse transform is a complex number in general. To ensure that $f(r)$ is real, the Fourier transform $F(q)$ must be such that the Friedel mates $F(-q)$ and $F(q)$ are complex conjugates of one another. Thus, $F(-q)$ has the same magnitude as $F(q)$ but they have the opposite phase, i.e., $\varphi(q) = -\varphi(q)$

$$F(-q) = |F(-q)|e^{i\phi(-q)} = F^*(q) = |F(q)|e^{-i\phi(q)}.$$

The equality of their magnitudes ensures that the Friedel mates have the same intensity $|F|^2$. This symmetry allows one to measure the full Fourier transform from only half the reciprocal space, e.g., by rotating the crystal slightly more than 180° instead of a full 360° revolution. In crystals with significant symmetry, even more reflections may have the same intensity (Bijvoet mates); in such cases, even less of the reciprocal space may need to be measured. In favorable cases of high symmetry, sometimes only 90° or even only 45° of data are required to completely explore the reciprocal space.

The Friedel-mate constraint can be derived from the definition of the inverse Fourier transform

$$f(r) = \int \frac{dq}{(2\pi)^3} F(q)e^{iq\cdot r} = \int \frac{dq}{(2\pi)^3} |F(q)|e^{i\phi(q)}e^{iq\cdot r}.$$

Since Euler's formula states that $e^{ix} = \cos(x) + i\sin(x)$, the inverse Fourier transform can be separated into a sum of a purely real part and a purely imaginary part.

$$f(r) = \int \frac{dq}{(2\pi)^3}|F(q)|e^{i(\phi+q\cdot r)} = \int \frac{dq}{(2\pi)^3}|F(q)|\cos(\phi+q\cdot r) + i\int \frac{dq}{(2\pi)^3}|F(q)|\sin(\phi+q\cdot r) = I_{cos} + iI_{sin}.$$

The function $f(r)$ is real if and only if the second integral I_{sin} is zero for all values of r. In turn, this is true if and only if the above constraint is satisfied

$$I_{sin} = \int \frac{dq}{(2\pi)^3}|F(q)|\sin(\phi+q\cdot r) = \int \frac{dq}{(2\pi)^3}|F(-q)|\sin(-\phi-q\cdot r) = -I_{sin},$$

since $I_{sin} = -I_{sin}$ implies that $I_{sin} = 0$.

Ewald's Sphere

Each X-ray diffraction image represents only a slice, a spherical slice of reciprocal space, as may be seen by the Ewald sphere construction. Both k_{out} and k_{in} have the same length, due to the elastic scattering, since the wavelength has not changed. Therefore, they may be represented as two radial vectors in a sphere in reciprocal space, which shows the values of q that are sampled in a given diffraction image. Since there is a slight spread in the incoming wavelengths of the incoming X-ray beam, the values of $|F(q)|$ can be measured only for q vectors located between the two spheres corresponding to those radii. Therefore, to obtain a full set of Fourier transform data, it is necessary to rotate the crystal through slightly more than 180°, or sometimes less if sufficient symmetry is present. A full 360° rotation is not needed because of a symmetry intrinsic to the Fourier transforms of real functions (such as the electron density), but "slightly more" than 180° is needed to cover all of reciprocal space within a given resolution because of the curvature of the Ewald sphere. In practice, the crystal is rocked by a small amount (0.25–1°) to incorporate reflections near the boundaries of the spherical Ewald's shells.

Patterson Function

A well-known result of Fourier transforms is the auto correlation theorem, which states that the autocorrelation $c(r)$ of a function $f(r)$ such as given below.

$$c(r) = \int dx f(x) f(x+r) = \int \frac{dq}{(2\pi)^3} C(q) e^{iq \cdot r}$$

The above Equation has a Fourier transform $C(q)$ that is the squared magnitude of $F(q)$.

$$C(q) = |F(q)|^2.$$

Therefore, the auto correlation function $c(r)$ of the electron density (also known as the *Patterson function*) can be computed directly from the reflection intensities, without computing the phases. In principle, this could be used to determine the crystal structure directly; however, it is difficult to realize in practice. The autocorrelation function corresponds to the distribution of vectors between atoms in the crystal; thus, a crystal of N atoms in its unit cell may have $N(N-1)$ peaks in its Patterson function. Given the inevitable errors in measuring the intensities, and the mathematical difficulties of reconstructing atomic positions from the interatomic vectors, this technique is rarely used to solve structures, except for the simplest crystals.

Advantages of a Crystal

In principle, an atomic structure could be determined from applying X-ray scattering to non-crystalline samples, even to a single molecule. However, crystals offer a much stronger signal due to their periodicity. A crystalline sample is by definition periodic; a crystal is composed of many unit cells repeated indefinitely in three independent directions. Such periodic systems have a Fourier transform that is concentrated at periodically repeating points in reciprocal space known as *Bragg peaks*; the Bragg peaks correspond to the reflection spots observed in the diffraction image. Since the amplitude at these reflections grows linearly with the number N of scatterers, the observed *intensity* of these spots should grow quadratically, like N^2. In other words, using a crystal concentrates the weak scattering of the individual unit cells into a much more powerful, coherent reflection that can be observed above the noise. This is an example of constructive interference.

In a liquid, powder or amorphous sample, molecules within that sample are in random orientations. Such samples have a continuous Fourier spectrum that uniformly spreads its amplitude thereby reducing the measured signal intensity, as is observed in SAXS. More importantly, the orientational information is lost. Although theoretically possible, it is experimentally difficult to obtain atomic-resolution structures of complicated, asymmetric molecules from such rotationally averaged data. An intermediate case is fiber diffraction in which the subunits are arranged periodically in at least one dimension.

Applications of X-ray Diffraction

X-ray diffraction has wide and various applications in the chemical, biochemical, physical, material and mineralogical sciences. Laue claimed in 1937 that the technique "has extended the power of observing minute structure ten thousand times beyond that given us by the microscope".X-ray diffraction is analogous to a microscope with atomic-level resolution which shows the atoms and their electron distribution.

X-ray diffraction, electron diffraction, and neutron diffraction give information about the structure of matter, crystalline and non-crystalline, at the atomic and molecular level. In addition, these methods may be applied in the study of properties of all materials, inorganic, organic or biological. Due to the importance and variety of applications of diffraction studies of crystals, many Nobel Prizes have been awarded for such studies.

X-ray Method for Investigation of Drugs

X-ray diffraction has been used for the identification of antibiotic drugs such as: eight β-lactam (ampicillin sodium, penicillin G procaine, cefalexin, ampicillin trihydrate, benzathine penicillin, benzylpenicillin sodium, cefotaxime sodium, Ceftriaxone sodium), three tetracycline (doxycycline hydrochloride, oxytetracycline dehydrate, tetracycline hydrochloride) and two macrolide (azithromycin, erythromycin estolate) antibiotic drugs. Each of these drugs has a unique X-Ray Diffraction (XRD) pattern that makes their identification possible.

X-ray Method for Investigation of Textile Fibers and Polymers

Forensic examination of any trace evidence is based upon Locard's exchange principle. This states that "every contact leaves a trace". In practice, even though a transfer of material has taken place, it may be impossible to detect, because the amount transferred is very small.

Textile fibers are a mixture of crystalline and amorphous substances. Therefore, the measurement of the degree of crystalline gives useful data in the characterization of fibers using X-ray diffractometry. It has been reported that X-ray diffraction was used to identify of a "crystalline" deposit which was found on a chair. The deposit was found to be amorphous, but the diffraction pattern present matched that of polymethylmethacrylate. Pyrolysis mass spectrometry later identified the deposit as polymethylcyanoacrylaon of Boin crystal parameters.

X-ray Method for Investigation of Bones

Heating or burning of bones causes recognizable changes in the bone mineral that can be detected using XRD techniques. During the first 15 minutes of heating at 500 °C or above, the bone crystals began to change. At higher temperatures, thickness and shape of crystals of bones appear stabilized, but when the samples were heated at a lower temperature or for a shorter period, XRD traces showed extreme changes in crystal parameters.

Integrated Circuits

X-ray diffraction has been demonstrated as a method for investigating the complex structure of integrated circuits.

Crystallization

Crystallization is the (natural or artificial) process by which a solid forms, where the atoms or molecules are highly organized into a structure known as a crystal. Some of the ways by which crystals form are precipitating from a solution, freezing, or more rarely deposition directly from a gas. Attributes of the resulting crystal depend largely on factors such as temperature, air pressure, and in the case of liquid crystals, time of fluid evaporation.

Crystallization occurs in two major steps. The first is nucleation, the appearance of a crystalline phase from either a supercooled liquid or a supersaturated solvent. The second step is known as crystal growth, which is the increase in the size of particles and leads to a crystal state. An important feature of this step is that loose particles form layers at the crystal's surface lodge themselves into open inconsistencies such as pores, cracks, etc.

The majority of minerals and organic molecules crystallize easily, and the resulting crystals are generally of good quality, i.e. without visible defects. However, larger biochemical particles, like proteins, are often difficult to crystallize. The ease with which molecules will crystallize strongly depends on the intensity of either atomic forces (in the case of mineral substances), intermolecular forces (organic and biochemical substances) or intramolecular forces (biochemical substances).

Crystallization is also a chemical solid–liquid separation technique, in which mass transfer of a solute from the liquid solution to a pure solid crystalline phase occurs. In chemical engineering, crystallization occurs in a crystallizer. Crystallization is therefore related to precipitation, although the result is not amorphous or disordered, but a crystal.

Process

The crystallization process consists of two major events, nucleation and crystal growth which are driven by thermodynamic properties as well as chemical properties. In crystallization Nucleation is the step where the solute molecules or atoms dispersed in the solvent start to gather into clusters, on the microscopic scale (elevating solute concentration in a small region), that become stable under the current operating conditions. These stable clusters constitute the nuclei. Therefore, the clusters need to reach a critical size in order to become stable nuclei. Such critical size is dictated by many different factors (temperature, supersaturation, etc.). It is at the stage of nucleation that the atoms or molecules arrange in a defined and periodic manner that defines the crystal structure — note that "crystal structure" is a special term that refers to the relative arrangement of the atoms or molecules, not the macroscopic properties of the crystal (size and shape), although those are a result of the internal crystal structure.

The crystal growth is the subsequent size increase of the nuclei that succeed in achieving the critical cluster size. Crystal growth is a dynamic process occurring in equilibrium where solute molecules or atoms precipitate out of solution, and dissolve back into solution. Supersaturation is one of the driving forces of crystallization, as the solubility of a species is an equilibrium process quantified by K_{sp}. Depending upon the conditions, either nucleation or growth may be predominant over the other, dictating crystal size.

Many compounds have the ability to crystallize with some having different crystal structures, a phenomenon called polymorphism. Each polymorph is in fact a different thermodynamic solid state and crystal polymorphs of the same compound exhibit different physical properties, such as dissolution rate, shape (angles between facets and facet growth rates), melting point, etc. For this reason, polymorphism is of major importance in industrial manufacture of crystalline products. Additionally, crystal phases can sometimes be interconverted by varying factors such as temperature, such as in the transformation of anatase to rutile phases of titanium dioxide.

In Nature

There are many examples of natural process that involve crystallization.

Geological time scale process examples include:

- Natural (mineral) crystal formation;
- Stalactite/stalagmite, rings formation.

Human time scale process examples include:

- Snow flakes formation;
- Honey crystallization (nearly all types of honey crystallize).

Snowflakes are a very well-known example, where subtle differences in crystal growth conditions result in different geometries.

Methods

Crystal formation can be divided into two types, where the first type of crystals are composed of a cation and anion, also known as a salt, such as sodium acetate. The second type of crystals are composed of uncharged species, for example menthol.

Crystallized honey.

Crystal formation can be achieved by various methods, such as: cooling, evaporation, addition of a second solvent to reduce the solubility of the solute (technique known as antisolvent or drown-out), solvent layering, sublimation, changing the cation or anion, as well as other methods.

The formation of a supersaturated solution does not guarantee crystal formation, and often a seed crystal or scratching the glass is required to form nucleation sites.

A typical laboratory technique for crystal formation is to dissolve the solid in a solution in which it is partially soluble, usually at high temperatures to obtain supersaturation. The hot mixture is then filtered to remove any insoluble impurities. The filtrate is allowed to slowly cool. Crystals that form are then filtered and washed with a solvent in which they are not soluble, but is miscible with the mother liquor. The process is then repeated to increase the purity in a technique known as recrystallization.

For biological molecules in which the solvent channels continue to be present to retain the three dimensional structure intact, microbatchcrystallization under oil and vapor diffusionmethods have been the common methods.

Typical Equipment

Equipment for the main industrial processes for crystallization:

Tank crystallizers- Tank crystallization is an old method still used in some specialized cases. Saturated solutions, in tank crystallization, are allowed to cool in open tanks. After a period of time the mother liquor is drained and the crystals removed. Nucleation and size of crystals are difficult to control. Typically, labor costs are very high.

Thermodynamic View

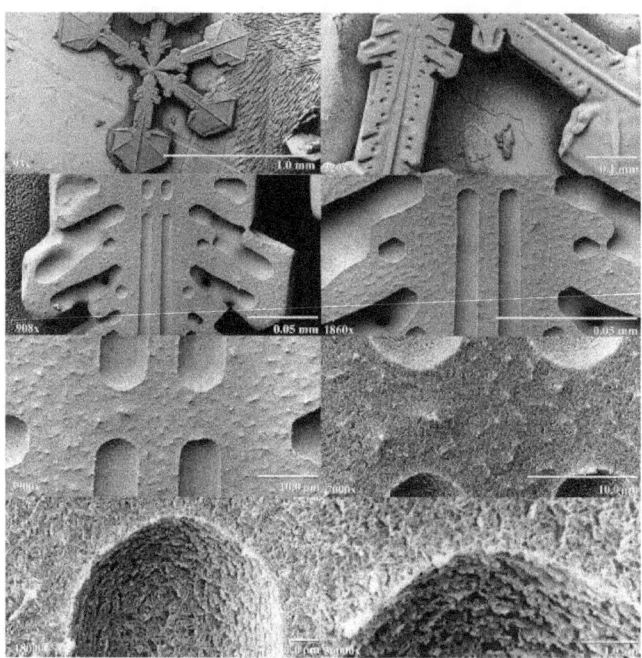

Low-temperature SEM magnification series for a snow crystal. The crystals are captured, stored, and sputter-coated with platinum at cryo-temperatures for imaging.

The crystallization process appears to violate the second principle of thermodynamics. Whereas most processes that yield more orderly results are achieved by applying heat, crystals usually form at lower temperatures—especially by supercooling. However, due to the release of the heat of fusion during crystallization, the entropy of the universe increases, thus this principle remains unaltered.

The molecules within a pure, perfect crystal, when heated by an external source, will become liquid. This occurs at a sharply defined temperature (different for each type of crystal). As it liquifies, the complicated architecture of the crystal collapses. Melting occurs because the entropy (S) gain in the system by spatial randomization of the molecules has overcome the enthalpy (H) loss due to breaking the crystal packing forces:

$$T(S_{\text{liquid}} - S_{\text{solid}}) > H_{\text{liquid}} - H_{\text{solid}},$$

$$G_{\text{liquid}} < G_{\text{solid}}.$$

Regarding crystals, there are no exceptions to this rule. Similarly, when the molten crystal is cooled, the molecules will return to their crystalline form once the temperature falls beyond the turning point. This is because the thermal randomization of the surroundings compensates for the loss of entropy that results from the reordering of molecules within the system. Such liquids that crystallize on cooling are the exception rather than the rule.

The nature of a crystallization process is governed by both thermodynamic and kinetic factors, which can make it highly variable and difficult to control. Factors such as impurity level, mixing regime, vessel design, and cooling profile can have a major impact on the size, number, and shape of crystals produced.

Dynamics

As mentioned above, a crystal is formed following a well-defined pattern, or structure, dictated by forces acting at the molecular level. As a consequence, during its formation process the crystal is in an environment where the solute concentration reaches a certain critical value, before changing status. Solid formation, impossible below the solubility threshold at the given temperature and pressure conditions, may then take place at a concentration higher than the theoretical solubility level. The difference between the actual value of the solute concentration at the crystallization limit and the theoretical (static) solubility threshold is called supersaturation and is a fundamental factor in crystallization.

Nucleation

Nucleation is the initiation of a phase change in a small region, such as the formation of a solid crystal from a liquid solution. It is a consequence of rapid local fluctuations on a molecular scale in a homogeneous phase that is in a state of metastable equilibrium. Total nucleation is the sum effect of two categories of nucleation – primary and secondary.

Primary Nucleation

Primary nucleation is the initial formation of a crystal where there are no other crystals present or where, if there are crystals present in the system, they do not have any influence on the process. This can occur in two conditions. The first is homogeneous nucleation, which is nucleation that is not influenced in any way by solids. These solids include the walls of the crystallizer vessel and particles of any foreign substance. The second category, then, is heterogeneous nucleation. This occurs when solid particles of foreign substances cause an increase in the rate of nucleation that would otherwise not be seen without the existence of these foreign particles. Homogeneous nucleation rarely occurs in practice due to the high energy necessary to begin nucleation without a solid surface to catalyse the nucleation.

Primary nucleation (both homogeneous and heterogeneous) has been modelled with the following:

$$B = \frac{dN}{dt} = k_n (c - c^*)^n,$$

where

> B is the number of nuclei formed per unit volume per unit time,
>
> N is the number of nuclei per unit volume,
>
> k_n is a rate constant,
>
> c is the instantaneous solute concentration,
>
> c^* is the solute concentration at saturation,
>
> $(c - c^*)$ is also known as supersaturation,
>
> n is an empirical exponent that can be as large as 10, but generally ranges between 3 and 4.

Secondary Nucleation

Secondary nucleation is the formation of nuclei attributable to the influence of the existing microscopic crystals in the magma. Simply put, secondary nucleation is when crystal growth is initiated with contact of other existing crystals or "seeds".The first type of known secondary crystallization is attributable to fluid shear, the other due to collisions between already existing crystals with either a solid surface of the crystallizer or with other crystals themselves. Fluid-shear nucleation occurs when liquid travels across a crystal at a high speed, sweeping away nuclei that would otherwise be incorporated into a crystal, causing the swept-away nuclei to become new crystals. Contact nucleation has been found to be the most effective and common method for nucleation. The benefits include the following:

- Low kinetic order and rate-proportional to supersaturation, allowing easy control without unstable operation.

- Occurs at low supersaturation, where growth rate is optimal for good quality.

- Low necessary energy at which crystals strike avoids the breaking of existing crystals into new crystals.

- The quantitative fundamentals have already been isolated and are being incorporated into practice.

The following model, although somewhat simplified, is often used to model secondary nucleation:

$$B = \frac{dN}{dt} = k_1 M_T^j (c - c^*)^b,$$

where

k_1 is a rate constant,

M_T is the suspension density,

j is an empirical exponent that can range up to 1.5, but is generally 1,

b is an empirical exponent that can range up to 5, but is generally 2.

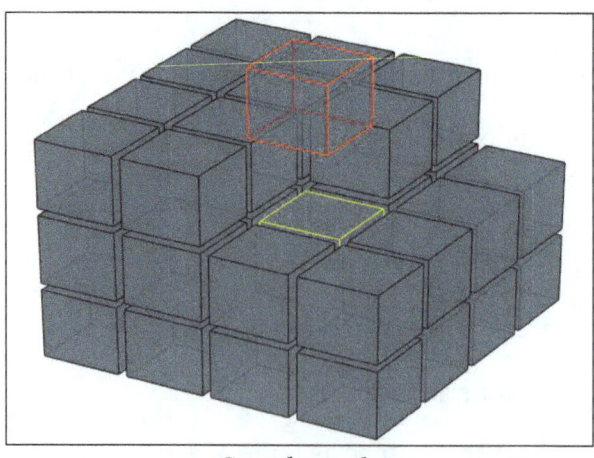

Crystal growth

Growth

Once the first small crystal, the nucleus, forms it acts as a convergence point (if unstable due to super-saturation) for molecules of solute touching – or adjacent to – the crystal so that it increases its own dimension in successive layers. The pattern of growth resembles the rings of an onion, as shown in the picture, where each colour indicates the same mass of solute; this mass creates increasingly thin layers due to the increasing surface area of the growing crystal. The supersaturated solute mass the original nucleus may capture in a time unit is called the growth rate expressed in $kg/(m^2 \cdot h)$, and is a constant specific to the process. Growth rate is influenced by several physical factors, such as surface tension of solution, pressure, temperature, relative crystal velocity in the solution, Reynolds number, and so forth.

The main values to control are therefore:

- Supersaturation value, as an index of the quantity of solute available for the growth of the crystal;

- Total crystal surface in unit fluid mass, as an index of the capability of the solute to fix onto the crystal;

- Retention time, as an index of the probability of a molecule of solute to come into contact with an existing crystal;

- Flow pattern, again as an index of the probability of a molecule of solute to come into contact with an existing crystal (higher in laminar flow, lower in turbulent flow, but the reverse applies to the probability of contact).

The first value is a consequence of the physical characteristics of the solution, while the others define a difference between a well- and poorly designed crystallizer.

Size Distribution

The appearance and size range of a crystalline product is extremely important in crystallization. If further processing of the crystals is desired, large crystals with uniform size are important for washing, filtering, transportation, and storage, because large crystals are easier to filter out of a solution than small crystals. Also, larger crystals have a smaller surface area to volume ratio, leading to a higher purity. This higher purity is due to less retention of mother liquor which contains impurities, and a smaller loss of yield when the crystals are washed to remove the mother liquor. The theoretical crystal size distribution can be estimated as a function of operating conditions with a fairly complicated mathematical process called population balance theory.

Main Crystallization Processes

Some of the important factors influencing solubility are:

- Concentration

- Temperature

- Polarity

- Ionic strength

So one may identify two main families of crystallization processes:

- Cooling crystallization

- Evaporative crystallization

This division is not really clear-cut, since hybrid systems exist, where cooling is performed through evaporation, thus obtaining at the same time a concentration of the solution.

A crystallization process often referred to in chemical engineering is the fractional crystallization. This is not a different process, rather a special application of one (or both) of the above.

Cooling Crystallization

Application

Most chemical compounds, dissolved in most solvents, show the so-called *direct* solubility that is, the solubility threshold increases with temperature.

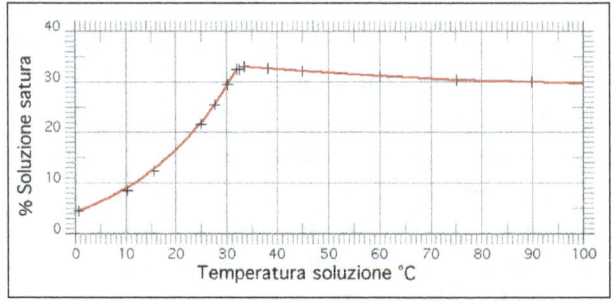

Solubility of the system $Na_2SO_4 - H_2O$

So, whenever the conditions are favourable, crystal formation results from simply cooling the solution. Here cooling is a relative term: austenite crystals in a steel form well above 1000 °C. An example of this crystallization process is the production of Glauber's salt, a crystalline form of sodium sulfate. In the diagram, where equilibrium temperature is on the x-axis and equilibrium concentration (as mass percent of solute in saturated solution) in y-axis, it is clear that sulfate solubility quickly decreases below 32.5 °C. Assuming a saturated solution at 30 °C, by cooling it to 0 °C (note that this is possible thanks to the freezing-point depression), the precipitation of a mass of sulfate occurs corresponding to the change in solubility from 29% (equilibrium value at 30 °C) to approximately 4.5% (at 0 °C) – actually a larger crystal mass is precipitated, since sulfate entrains hydration water, and this has the side effect of increasing the final concentration.

There are limitations in the use of cooling crystallization:

- Many solutes precipitate in hydrate form at low temperatures: in the previous example this is acceptable, and even useful, but it may be detrimental when, for example, the mass of water of hydration to reach a stable hydrate crystallization form is more than the available water: a single block of hydrate solute will be formed – this occurs in the case of calcium chloride);

- Maximum supersaturation will take place in the coldest points. These may be the heat exchanger tubes which are sensitive to scaling, and heat exchange may be greatly reduced or discontinued;

- A decrease in temperature usually implies an increase of the viscosity of a solution. Too high a viscosity may give hydraulic problems, and the laminar flow thus created may affect the crystallization dynamics;

- It is not applicable to compounds having *reverse* solubility, a term to indicate that solubility increases with temperature decrease (an example occurs with sodium sulfate where solubility is reversed above 32.5 °C).

Cooling Crystallizers

Vertical cooling crystallizer in a beet sugar factory.

The simplest cooling crystallizers are tanks provided with a mixer for internal circulation, where temperature decrease is obtained by heat exchange with an intermediate fluid circulating in a jacket. These simple machines are used in batch processes, as in processing of pharmaceuticals and are prone to scaling. Batch processes normally provide a relatively variable quality of product along the batch.

The Swenson-Walker crystallizer is a model, specifically conceived by Swenson Co. around 1920, having a semicylindric horizontal hollow trough in which a hollow screw conveyor or some hollow discs, in which a refrigerating fluid is circulated, plunge during rotation on a longitudinal axis. The refrigerating fluid is sometimes also circulated in a jacket around the trough. Crystals precipitate on the cold surfaces of the screw/discs, from which they are removed by scrapers and settle on the bottom of the trough. The screw, if provided, pushes the slurry towards a discharge port.

A common practice is to cool the solutions by flash evaporation: when a liquid at a given T_0 temperature is transferred in a chamber at a pressure P_1 such that the liquid saturation temperature T_1 at P_1 is lower than T_0, the liquid will release heat according to the temperature difference and a quantity of solvent, whose total latent heat of vaporization equals the difference in enthalpy. In simple words, the liquid is cooled by evaporating a part of it.

In the sugar industry, vertical cooling crystallizers are used to exhaust the molasses in the last crystallization stage downstream of vacuum pans, prior to centrifugation. The massecuite enters the crystallizers at the top, and cooling water is pumped through pipes in counterflow.

Evaporative Crystallization

Another option is to obtain, at an approximately constant temperature, the precipitation of the crystals by increasing the solute concentration above the solubility threshold. To obtain this, the solute/solvent mass ratio is increased using the technique of evaporation. This process is insensitive to change in temperature (as long as hydration state remains unchanged).

All considerations on control of crystallization parameters are the same as for the cooling models.

Evaporative Crystallizers

Most industrial crystallizers are of the evaporative type, such as the very large sodium chloride and sucrose units, whose production accounts for more than 50% of the total world production of crystals. The most common type is the forced circulation (FC) model. A pumping device (a pump or an axial flow mixer) keeps the crystal slurry in homogeneous suspension throughout the tank, including the exchange surfaces; by controlling pump flow, control of the contact time of the crystal mass with the supersaturated solution is achieved, together with reasonable velocities at the exchange surfaces. The Oslo, mentioned above, is a refining of the evaporative forced circulation crystallizer, now equipped with a large crystals settling zone to increase the retention time (usually low in the FC) and to roughly separate heavy slurry zones from clear liquid. Evaporative crystallizers tend to yield larger average crystal size and narrows the crystal size distribution curve.

DTB Crystallizer

Whichever the form of the crystallizer, to achieve an effective process control it is important to control the retention time and the crystal mass, to obtain the optimum conditions in terms of crystal specific surface and the fastest possible growth. This is achieved by a separation – to put it simply – of the crystals from the liquid mass, in order to manage the two flows in a different way. The practical way is to perform a gravity settling to be able to extract (and possibly recycle separately) the (almost) clear liquid, while managing the mass flow around the crystallizer to obtain a precise slurry density elsewhere. A typical example is the DTB (*Draft Tube and Baffle*) crystallizer, an idea of Richard Chisum Bennett at the end of the 1950s. The DTB crystallizer has an internal circulator, typically an axial flow mixer – yellow – pushing upwards in a draft tube while outside the crystallizer there is a settling area in an annulus; in it the exhaust solution moves upwards at a very low velocity, so that large crystals settle – and return to the main circulation – while only the fines, below a given grain size are extracted and eventually destroyed by increasing or decreasing temperature, thus creating additional supersaturation. A quasi-perfect control of all parameters is achieved as DTF crystallizers offer superior control over crystal size and characteristics. This crystallizer, and the derivative models (Krystal, CSC, etc.) could be the ultimate solution if not for a major limitation in the evaporative capacity, due to the limited diameter of the vapour head and the relatively low external circulation not allowing large amounts of energy to be supplied to the system.

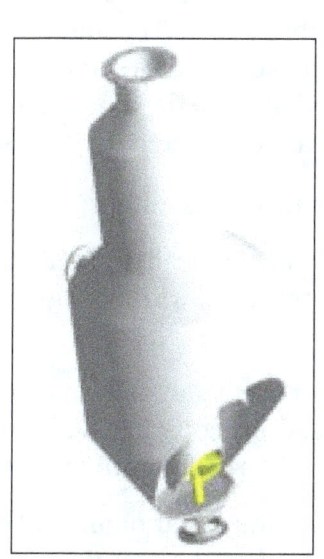

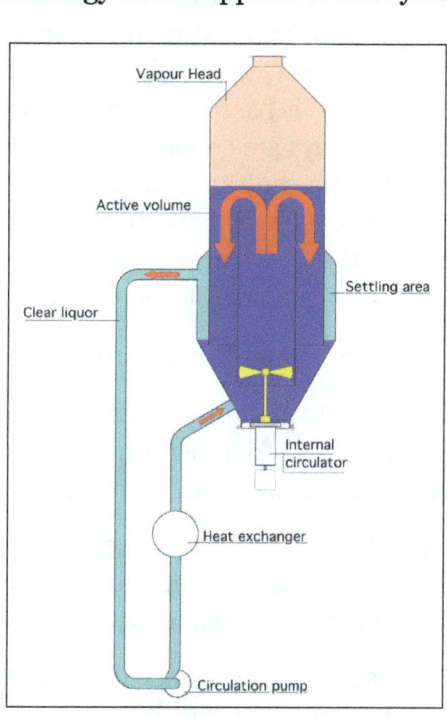

DTB Crystallizer.   Schematic of DTB

X-ray Crystallography

X-ray crystallography is a tool used for identifying the atomic and molecular structure of a crystal, in which the crystalline atoms cause a beam of incident X-rays to diffract into many specific directions. By measuring the angles and intensities of these diffracted beams, a crystallographer can produce a three-dimensional picture of the density of electrons within the crystal. From this electron density, the mean position of the atoms, chemical bonds and their disorder in the crystal can be determined. For single crystals of sufficient purity and regularity, X-ray diffraction data can determine the mean chemical bond lengths and angles to within a few thousandths of an angstrom and to within a few tenths of a degree, respectively. The atoms in a crystal are not static, but oscillate about their mean positions, usually by less than a few tenths of an angstrom. X-ray crystallography allows measuring the size of these oscillations and also helps in relating the structure and function of many biological molecules such as vitamins, drugs, proteins and nucleic acids. X-ray crystal structures can also account for unusual electronic or elastic properties of a material, shed light on chemical interactions and processes, or serve as the basis for designing pharmaceuticals against diseases. Crystallographers study diverse substances from living cells to superconductors, from protein molecule to ceramics.

Crystallographic methods depend on analysis of the diffraction patterns of a sample targeted by a beam X-rays, electrons or neutrons. This is based on the wave properties of the particles. X-rays interact with the spatial distribution of electrons in the sample. Electrons are charged particles and therefore interact with the total charge distribution of both the atomic nuclei and the electrons of the sample. Neutrons are scattered by the atomic nuclei through the strong nuclear forces, but the magnetic moment of neutrons is non-zero. Therefore they are also scattered by magnetic fields.

Among these three methods X-ray diffraction is the most common technique used for the study of crystal structures and atomic spacing. Max von Laue, in 1912, discovered that crystalline substances act as three-dimensional diffraction gratings for X-ray wavelengths similar to the spacing of planes in a crystal lattice. X-ray diffraction is based on constructive interference of monochromatic X-rays and a crystalline sample. These X-rays are generated by a cathode ray tube, filtered to produce monochromatic radiation, collimated to concentrate, and directed towards the sample. The interaction of the incident rays with the sample produces constructive interference when conditions satisfy Bragg's Law ($n\lambda = 2d \sin \theta$). This law relates the wavelength of electromagnetic radiation to the diffraction angle and the lattice spacing in a crystalline sample. These diffracted X-rays are then detected, processed and counted.

The technique of single-crystal X-ray crystallography has three basic steps:

1. The first step is to obtain an adequate crystal of size typically larger than 0.1 mm in all dimensions, pure in composition and regular in structure, with no significant internal imperfections such as cracks or twinning.

2. In the second step, the crystal is exposed to an intense beam of X-rays, usually of a single wavelength producing the regular pattern of reflections covering slightly more than half a full rotation of the crystal and typically containing tens of thousands of reflections.

3. In the third step, the intensity data are combined computationally with complementary chemical information to produce and refine a model of the atomic arrangement called a crystal structure which is usually stored in a public database.

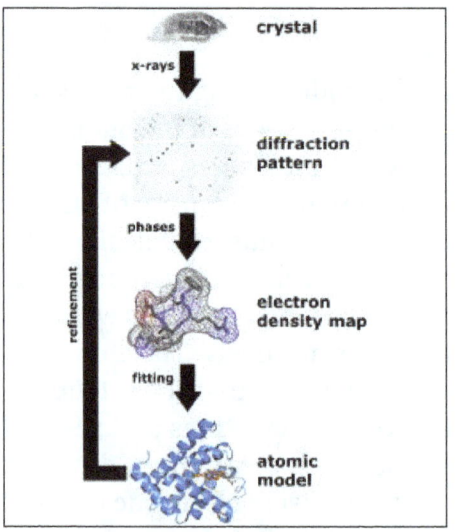

Workflow for solving the structure of a molecule.

Two limiting cases of X-ray crystallography are "small-molecule" and "macromolecular" crystallography. Small-molecule crystallography typically involves crystals with fewer than 100 atoms in their asymmetric unit; such crystal structures are usually so well resolved that the atoms can be discerned as isolated "blobs" of electron density. Macromolecular crystallography often involves tens of thousands of atoms in the unit cell. Such crystal structures are generally less well-resolved (more "smeared out"); the atoms and chemical bonds appear as tubes of electron density, rather than as isolated atoms.

Crystal Structure Determination

Crystal structure determination of a compound required to synthesize a diffraction quality crystal, collecting intensity data using X-ray diffractometer, elucidating E-map, identifying the atomic positions, solving structure to identify the elements and refining the structure to measure closeness between the proposed and the derived structures.

Growing Crystals

Although crystals can be grown by depositions from any sort of fluid phase for most organic and inorganic compounds, the most practical method such as slow evaporation, slow cooling and diffusion involves crystallization from a solution in a suitable solvent. Among these three methods slow evaporation is most suitable for small molecules.

Slow evaporation often deposit crystals as a microcrystalline crust on the walls of the container just at the surface of the solution. As the solvent evaporates, the solution recedes, leaving the crust in a position where it is not effective in inducing good crystal growth.

Intensity Data Collection

A crystal of suitable size is used to collect the intensity data using 'Brucker AXS KAPPA APEX-II and SMART CCD area detector. The crystal data used in the present research work were collected using 'BRUKER AXS KAPPA APEX-II CCD diffractometer'. The diffractometer is equipped with MoKα (0.71073Å) radiation and the high Bragg angle reflections are collected at room temperature.

BRUKER AXS KAPPA APEX -II CCD Diffractometer.

Nowadays, crystals of small molecule are typically attached with oil or glue to a glass fiber or a loop, which is made of nylon or plastic and attached to a solid rod. The capillary or loop is mounted on a goniometer, which allows it to be positioned accurately within the X-ray beam and rotated. Since both the crystal and the beam are often very small, the crystal must be centered within the beam to within ~25 micrometers accuracy, which is aided by a camera focused on the crystal. The most common type of goniometer is the "kappa goniometer", which offers three angles of rotation: the ω angle, which rotates about an axis perpendicular to the beam; the κ angle, about an axis at ~50° to the ω axis; and, finally, the φ angle about the loop/capillary axis. When the κ angle is zero, the ω and φ axes are aligned. The κ rotation allows for convenient mounting of the crystal, and the oscillations carried out during data collection involve the ω axis only.

The unit cell parameters are determined from 36 frames (0.5° Phi-Scan) measured from three different Crystallographic Zones by using the method of difference vectors. The intensity data is collected with an average four-fold redundancy per reflection and optimum resolution (0.75Å). The intensity data collection, frames integration, Lorentz polarization (Lp) correction and decay correction were carried out using SAINT software. Empirical absorption correction (multi-scan) is performed using SADABS program. All the intensities were corrected for variable scan speed, background and attenuation using the relation.

$$I_{raw} = f \left[N_c - 2 \left(L_b + R_b \right) \right] NPI$$

where,

I_{raw} - Relative intensity;

N_c - Peak Count;

L_b and R_b - Left and right background counts, respectively;

NPI - Scan speed parameter;

f - Attenuation factor.

Data Reduction

The preliminary manipulation of the intensities and their correction to a corrected form is referred as data reduction i.e conversion of raw data into a usable data. Geometrical corrections can include changes in the incident X-ray beam intensity, the scattering power of the crystal, absorption corrections that depend on the size and dimensions of the crystal and the calculation of path lengths. A statistical analysis of the complete unique data set can provide an indication of the presence or absence of some symmetry elements.

The space group of the crystal was determined from the systematic absence of the reflections and by intensity statistics. If space group ambiguity arises then the contents of the unit cell, the number of molecules present in the cell, the distribution of intensities and other relevant details are analyzed in depth.

The structure factors are the quantities that are used in the calculation of electron density maps from which the positions of atoms can be determined. Therefore the intensities are converted in to "observed " structure amplitudes $|F_o| = |F_{observed}|$ by a data reduction program.

The relationship between $|F_o|$ and I depends on a number of factors, mainly geometric, that are related to the individual reflection and to the apparatus used to measure its intensity. The observed structure factor for each reflection is obtained using the equation:

$$\left| F_{hk1} \right| = K \left(I_{hkl} / L_p \right)^{1/2}$$

K - Scaling Factor

L - Lorentz Factor

p - Polarization Factor

Lorentz Factor

The Lorentz factor depends both on the Bragg angle and on the diffraction geometry. It is proportional to the time of reflections permitted to each reflection, or inversely proportional to the velocity with which the plane passes through the condition of reflection.

Lorentz factor is given by $L = 1 / sin2\theta$

Polarization Factor

The polarization term "p" arises because of the nature of the X-ray beam and the manner in which its reflection efficiency varies with the reflection angle. The direction of polarization of an X-ray photon can change as a result of scattering/diffraction. When the change is maximal or when there is no change the two extreme cases to consider are;

- The component of polarization resolved along the diffracted direction is reduced by the cosine of the scattering angle (2θ) and so the reduction in intensity is $cos^2 2\theta$ (since intensity is proportional to the square of the amplitude).

- The component of polarization is clearly unaffected, and therefore unchanged, by the diffraction process so that the reduction factor is "1". Polarization factor is given by:

$$p = \left(1 + cos^2 2\theta \right) / 2$$

X-Ray Absorption

Crystals absorb X-rays thereby causing a reduction in the intensities of the reflection data. Each reflection is affected differently by absorption, because the absorption depends on the path length of the X-rays through the crystal, and this varies as the orientation of the crystal is changed. The amount of absorption is a function of the path length through the crystal and the absorption coefficients of the atoms making up the crystal.

X-rays are absorbed as they pass through materials according to the exponential law:

$$I = I_o e^{-\mu t}$$

where,

I$_o$ - the incident intensity

I - intensity after travelling a distance t through the material

μ - the absorption coefficient

Absorption can affect the accuracy of the bond distances and angles and cause unusual shaped thermal ellipsoids of the atoms, inflate the agreement factors, and in some cases inhibit the ability to solve the structure.

Temperature Effects

In X-ray diffraction spectra, it is assumed that the crystal structure is a static cone, i.e., it can be thought as a periodic pattern of stationary atoms. As the temperature rises the atoms execute increased excursions from their average position and the electrons of each atom sweeps out a larger average volume than they would occupy if the atom were at rest. This causes the effective fT curves of the atoms to fall off more rapidly with sin θ/λ, than for the same atom at rest. Thus a Debye-Waller temperature correction is applied to the scattering factor, which is given by:

$$f_T = f \exp{(-B \sin^2 \theta / \lambda^2)}$$

where,

T - The temperature;

B - The thermal factor and its value is $8\pi^2 u^2$ (units of Å^2);

<u^2 > - The root mean square displacement of the atom from its average position.

Structure Solution

The structure solution can be obtained by any of the methods that determine the correct phases without any ambiguities. The data collected from a diffraction experiment consists of diffraction 'spot' and the intensity of each diffraction 'spot' is proportional to the square of the structure factor amplitude. The structure factor is a complex number containing information relating to both the amplitude and phase of a wave. In order to obtain an interpretable electron density map, both amplitude and phase must be known (an electron density map allows a crystallographer to build a starting model of the molecule). The phase cannot be directly recorded during a diffraction experiment: this is known as the phase problem.

In X-ray crystallography, the diffraction data when properly assembled gives the amplitude of the 3D Fourier transform of the molecule's electron density in the unit cell. If the phases are known, the electron density can be simply obtained by Fourier synthesis. In order to obtain atomic positions of the molecule, intensities are converted into structure factors. Structure factor is the resultant of N waves scattered by N atoms in the unit cell. The general expression for the structure factor is:

$$F_{hkl} = \sum_{j=1}^{N} f_j \exp[2\pi i(hx_j + Ky_j + lz_j)]$$

Here, xj, yj, zj are fractional co-ordinates of jth atom;

N - total number of atoms in the unit cell;

f_j - atomic scattering factor of the jth atom.

Since F_{hkl} is a complex quantity, it can be written as:

$$F_{hkl} = |F_{hkl}| e^{ij}_{hkl}$$

where, $|F_{hkl}|$ is the structure factor amplitude of a Bragg reflection h, k, l and φ_{hkl} is the phase of the reflection h, k, l.

The structure factor can be expressed in terms of the integral of electrondensity. The Fourier transform of the structure factor yields electron-density. The three dimensional structure elucidation of any molecule is to obtain the three dimensional fractional coordinates (x, y, z) of atoms constituting the unit cell. The electron density maxima correspond to the atomic site and its value in the unit cell can be written as,

$$\rho(x,y,z) = (1/V) \sum_{h} \sum_{k} \sum_{l} F_{hkl} e^{-2\pi i(hx+ky+lz)}$$

where, V is the volume of the unit cell.

Method of Solving Phase Problem

There are various methods to solve the phase problem. Some of them are:

- Patterson Methods
- Direct Methods
- Multiple Isomorphous Replacement
- Multi wave length Anomalous Dispersion(MAD)
- Molecular Replacement

Of the developed methods direct methods is the most commonly used method for solving the structure of small molecules.

Patterson Methods

The Patterson function gives a large value in a position which corresponds to inter atomic vectors. This method can be applied only when the crystal contains heavy atoms or when a significant fraction of the structure is already known. The Patterson function has been an effective tool for solving small molecules; however, its usefulness falls quickly as the number of atoms increases.

Direct Methods

In this method the missing phase information may be reconstructed directly from mathematical relationships between the structure factors. Since the phases come directly from the observed diffraction pattern, these methods are referred to as "direct methods". A direct methods calculation might then proceed as follows. Phases are chosen for a few strong reflections and then phases for other reflections are generated using phase relationships among strong reflections. Once enough phases have been calculated, the electron density may be calculated and can be interpreted in terms of atomic positions. Because of the development of computers, the direct method is now the most useful technique for solving the phase problem.

A real crystal with continuous electron density is replaced by an idealized one consisting of N, non-vibrating point atoms. The electron density is assumed to be positive everywhere. For easier mathematical calculation equal atom structure was assumed. The ordinary structure factor given below corresponding to spherical atom has to be replaced for a corresponding point atom.

$$F_{hkl} = \sum_{j=1}^{N} f_j e^{2\pi i(hx_j + ky_j + lz_j)}$$

This quantity, the normalized structure factor amplitude ($|Ehkl|$) can be related to $|Fhkl|$ as:

$$\left| E_{hkl} \right|^2 = \left| F_{hkl} \right|^2 / (\varepsilon \sum_{j=1}^{N} f_j^{\,2})$$

Here ε is a constant (depends on the parity of reflections) and f_j is the thermal factor corrected atomic scattering factor.

Structure Refinement

Refinement is the process of iterative alteration of the molecular model with the goal to maximize its compliance with the diffraction data. For every atom in the model that is located on a general position in the unit cell, there are three atomic coordinates and six anisotropic displacement parameters to be refined. A second group of parameters, which are usually not refined in the least-squares procedure but adjusted by the crystallographer, are the atom types and site occupancy factors (SOF). A stable and reliable refinement requires a minimum number of observations per refined parameter, and the IUCr community currently recommends a minimum datato-parameter ratio of eight for non-centro symmetric structures and 10 for centrosymmetric structures. Excellent descriptions of the use of constraints and restraints in crystal structure refinements have been given by Watkin. Constraints are equations which assign fixed numerical values to certain parameters, hence reducing the number of independent parameters to be refined. When the data to-parameter ratio is low, or when correlations among certain parameters occur restraints can become essential.

A full-matrix least-squares refinement method is used in small molecular structure refinement using the program SHELXL-97. The least-squares refinement method uses the squares of the differences between the observed and calculated structure factor amplitudes and adjusts the parameters so that the disagreement is a minimum. The refinement of $F_o{}^2$ using all the data provides a good result for weakly diffracting crystals and for pseudo symmetry problems.

The residual factor or reliability index defining the correctness of the model is given by:

$$R = \{\Sigma|F_o| - |F_C|\} \ / \ \Sigma|Fo|$$

where,

   $|F_o|$ - observed structure factor amplitude;

   $|F_C|$ - calculated structure factor amplitude; and

   R-value should be minimum for the accurate model.

A weighting scheme is to be applied at the end of refinement procedure and the weighted factor is given by:

$$wR = \left\{ \left[\sum w \left(|F_o|^2 - |F_c|^2 \right)^2 \right] \ / \sum w \left(|F_o|^2 \right)^2 \right\}^{1/2}$$

where, each reflection has its own weight "w". wR will be nearly 3 to 4 times that of R.

A useful index often output by least squares refinement program is the goodness of fit, sometimes also termed the standard deviation of an observation of unit weight. The Goodness of Fit is always based on F^2

$$GooF = S = \left[\frac{\sum w(F_o^2 - F_c^2)^2}{(n-p)} \right]^{\frac{1}{2}}$$

Here, n is the number of reflections and p is the total number of parameters refined.

$$w = 1 / \left[\sigma^2 \left(Fo^2 \right) + \left(aP \right)^2 + bP \right]$$

Here, a and b are the constants and $P = \left[2Fc^2 + Max \left(Fo^2, 0 \right) / 3 \right]$.

Calculation of Geometrical Parameters

The most important anxiety in the determination of a crystal structure by diffraction is the arrangement of atoms in three-dimensional space. From the symmetry and coordinates of the atoms within the unit cell one can determine the coordination and bonding of the individual atoms. In order to do this, one have to calculate inter atomic distances, inter atomic angles, torsion angles, distances between planes of atoms, and so on.

Bond Length

Bond length is a measurable distance between atoms covalently bonded together. The length of the bond between two atoms is approximately the sum of the covalent radii of the two atoms. Therefore, bond length increases in the following order: triple bond < double bond < single bond. For a triclinic lattice, the distance between two points in fractional coordinates (x_1, y_1, z_1) and (x_2, y_2, z_2) is given by the law of cosines in three dimensions as:

$$L = \{(\Delta xa)^2 + (\Delta yb)^2 + (\Delta zc)^2 - 2ab\,\Delta x\,\Delta y\cos\gamma - 2ac\,\Delta x\,\Delta z\cos\beta - 2ba\,\Delta y\,\Delta z\cos\alpha\}^{1/2}$$

Where, a, b, c, α, β, γ are the unit-cell parameters.

The above equation is used to calculate the bond length. The bond length between "i" and "j" atom is represented as r_{ij} and shown in figure.

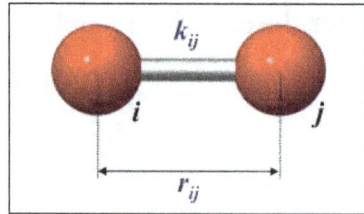

Bond distance

Bond Angle

The average angle between the orbitals of the central atom containing the bonding electron pairs in the molecule is known as bond angle between the atoms. This gives an idea about the distribution of orbital around the central atom in a molecule and also determines the shape of a molecule. By using the trigonometric cosine rule the angle between any three atoms is calculated as follows:

$$Cos\theta_{ij} = \left(A^2 + B^2 - C^2\right)/2AB$$

where, A, B, C are the lengths of the sides of the triangle ABC, and θ_{ij} is the angle ∠ A-B-C.

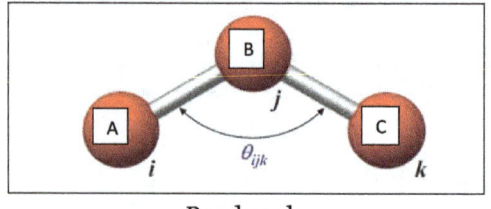

Bond angle

Torsion Angle

In addition to bond lengths and bond angles, an important parameter in describing the conformation of many organic and bioorganic molecules is the torsion angle. Torsion angles are dihedral angles, which are defined by 4 points in space. A dihedral angle or torsion angle is the angle between two planes. It defines the conformations around rotatable bonds. The dihedral angle changes only with the distance between the first and fourth atoms; the other inter atomic distances are controlled by the chemical bond lengths and bond angles. Its value ranges from -180 to +180 degrees. The torsion angle is considered to be positive if a clockwise rotation is performed with the molecule and it will be negative when an anti-clockwise rotation is performed with the molecule in its plane. The torsion angle is:

$$\phi_{ijkl} = -a \tan\left(\frac{\sin \phi_{ikjl}}{\cos \phi_{ikjl}}\right)$$

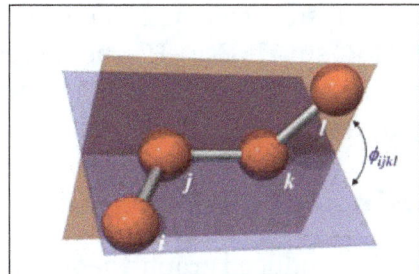

Torsion angle

Molecular Interactions

Molecular Interactions are forces, either attractive or repulsive, between molecules. They are important in diverse fields of protein folding, drug design, separation technologies, etc. The rationalization of crystal structures stems from a basic understanding of the nature of molecular interactions, which provide pathways for recognition and packing in the lattice. Interactions between two or more molecules are called intermolecular interactions, while the interactions between the atoms within a molecule are called intra molecular interactions. Intermolecular interactions are of enormous importance as they provide for energy minimization in the framework restricted by the crystal symmetry. Intermolecular interactions also provide useful guidelines for structure–activity correlations. In this context, hydrogen bonds play a major role since they cover the entire energy range between the covalent bonds and the van der Waals interactions. Intermolecular interactions range from the strong, long-distance electrical attractions and repulsions between ions to the relatively weak dispersion forces. The various types of interactions are classified as: Ion-ion, ion-dipole, dipole–dipole, ion-induced dipole, dipole-induced dipole and dispersion forces interactions.

Van Der Waals Interactions

Van der Waals forces are due to the induced electrical interactions between two or more atoms or molecules that are very close to each other. Van der Waals interaction is the weakest of all intermolecular attractions between molecules. These forces are due to the electrostatic attraction between the nucleus of one atom and the electrons of the other.

Hydrogen Bond

Hydrogen bonds are most important in biological systems and also of all directional intermolecular interactions. It is used in determining molecular conformation, molecular aggregation, and funtion of vast number of chemical systems ranging from inorganic to biological. Hydrogen bonding is the specific type of non-bonded interaction between two electronegative atoms, where hydrogen atom is covalently bonded to one of them. The hydrogen bond is represented as D-H...A, where D is the donor and A is the acceptor. The hydrogen bonds are highly directional and for an ideal case the D-H...A angle should be 180°. Hydrogen bond energy ranges from 15-40 Kcal/mol for strong bonds, 4-15 Kcal/mol for moderate bonds and 1-4 Kcal/mol for weak bonds. The hydrogen bond is a long-range interaction, and thus the existence for a donor group to be bonded to more than one acceptor at a time is called a bifurcated bond. The distance between the donor and acceptor is shorter by at least 0.2Å than the sum of their van der Walls radii for the hydrogen bond present in the crystal structure. The proton donor capacity of a C-H group depends on the hybridization [Csp-H>C sp²-H>Csp³-H] and increases with the number of adjacent withdrawing groups. Hydrogen bonds are investigated by a variety of experimental techniques, such as, Neutron diffraction, X-ray diffraction, NMR, IR and other spectroscopic techniques.

Hydrogen Bonded Motifs using Graph-Sets

Etter, Bernstein and co-workers have introduced graph-theory for describing and analyzing hydrogen bond networks in three-dimensional solids. A graph-set descriptor is designated as $G_D^A(n)$,

where,

G = Graph set designator C/R/D/S;

C - (chain)/R (ring)/D (dimer)/S (intramolecular hydrogen bonds);

D - Number of donor atoms;

A - Number of acceptor atoms;

n - Total number of atoms present in the hydrogen bonded motif.

All hydrogen bonding patterns can be described in terms of chains (C), rings (R), dimer (D) and intramolecular hydrogen bonds (S). The intermolecular interactions and hydrogen bond motifs that occur repeatedly in crystal structures are called supramolecular synthons. Synthons are the recognition motifs between building blocks that can be used to propagate networks or supramolecular assemblies.

References

- Snigirev, A. (2007). "Two-step hard X-ray focusing combining Fresnel zone plate and single-bounce ellipsoidal capillary". Journal of Synchrotron Radiation. 14 (Pt 4): 326–330. Doi:10.1107/S0909049507025174. PMID 17587657

- Crystallography, chemistry-careers, college-to-career, careers: acs.org, Retrieved 30 July, 2019

- Prince, E. (2006). International Tables for Crystallography Vol. C: Mathematical, Physical and Chemical Tables. Wiley. ISBN 978-1-4020-4969-9

- Stenos-law, science: britannica.com, Retrieved 31 August, 2019

- Patti Wigington (31 August 2016). "Using Crystals and Gemstones in Magic". About.com. Retrieved 14 November 2016

- Morphological-crystallography/Law-of-Rational-Indices, Crystallography, Supplemental-Modules-(Inorganic-Chemistry), Inorganic-Chemistry, Bookshelves: chem.libretexts.org, Retrieved 3 April, 2019

- Van der Put; Paul J. (2001). The inorganic chemistry of materials: how to make things out of elements. Academic Press. Pp. 123–126. ISBN 978-0-12-352651-9

- Rupp B, Wang J (November 2004). "Predictive models for protein crystallization". Methods. 34 (3): 390–407. Doi:10.1016/j.ymeth.2004.03.031. PMID 15325656

- Brown, Dwayne (October 30, 2012). "NASA Rover's First Soil Studies Help Fingerprint Martian Minerals". NASA. Retrieved October 31, 2012

PERMISSIONS

INDEX

www.ingramcontent.com/pod-product-compliance
Lightning Source LLC
Chambersburg PA
CBHW080404190526
45161CB00003B/130